Community College Mathematics

Community College Mathematics

Past, Present, and Future

Brian Cafarella

CRC Press
Taylor & Francis Group
Boca Raton London New York

CRC Press is an imprint of the
Taylor & Francis Group, an **informa** business

A CHAPMAN & HALL BOOK

First edition published 2022
by CRC Press
6000 Broken Sound Parkway NW, Suite 300, Boca Raton, FL 33487-2742

and by CRC Press
2 Park Square, Milton Park, Abingdon, Oxon, OX14 4RN

CRC Press is an imprint of Taylor & Francis Group, LLC

ISBN: 9781032262321 (hbk)
ISBN: 9781032262338 (pbk)
ISBN: 9781003287254 (ebk)

DOI: 10.1201/9781003287254

Typeset in Palatino
By codeMantra

This is for my beautiful wife, Lisa, and my amazing son,

Gavin, who make every day wonderful.

This is for my parents, Margaret and John, who always believed in me.

This is for my wonderful in-laws, Ethel and Roger,

who help our family in so many ways.

This is for the educational leadership faculty at the University of Dayton,

especially Dr. Michele Welkener and the late Dr. Theodore Kowalksi,

who sparked and nurtured my interest in academic writing.

A special thank you to Dr. Steven Johnson who

provided me with the idea for this book.

Contents

Preface...xvii
Author... xxi

1. **Higher Education: From the Elite to the Mass to the Universal** 1
   The Inception of American Higher Education 1
   The Start of Mathematics in Higher Education.......................................2
   The Changing Landscape of Higher Education..2
   The Progressive Movement..4
   The Two-Year Junior College...5
   A Focus on Math and a Greater Need for Developmental Math.............6
   The Rise of the Community College...7
   The Uniqueness of the Community College Model 8
   Summary...9
   References 10

2. **The 1970s: Mathematics and the Community College Become Acquainted**..13
   The Participants 13
   Professor Ballard...13
   Professor Mitchell...14
   Professor Milacki 14
   Professor Morgan..14
   Professor DeLeon..14
   Professor Fenimore...15
   Professor Wallace..15
   Professional Development...15
   Placement Testing 16
   Course Structure..19
   Sisco Community College (Quarter System) 19
       Basic Mathematics (Developmental Math—No Credit) 19
       Algebra 1 (Credit-Bearing Course) 19
       Algebra 2 (Credit-Bearing Course) 19
       College Algebra (Credit-Bearing Course)..19
       Statistics (Credit-Bearing Course)...19
   Telford Community College (Quarter System).......................................20
       Fundamentals of Mathematics (Developmental Math).....................20
       Elementary Algebra (Credit-Bearing Course in
       the Math Department)..20
       Intermediate Algebra (Credit-Bearing Course in
       the Math Department)..20

College Algebra (Credit-Bearing Course in the Math Department) 20
Statistics (Credit-Bearing Course in the Math Department).............. 20
Bordi Community College (Quarter System) 20
Basic Concepts of Mathematics (Developmental Math)..................... 20
Fundamentals of Algebra 1 (Credit-Bearing Course) 20
Fundamentals of Algebra 2 (Credit-Bearing Course) 21
College Algebra (Credit-Bearing Course)... 21
Statistics (Credit-Bearing Course)... 21
Habyan Community College (Quarter System) 21
Basic Mathematics (Credit-Bearing Course).. 21
Elementary Algebra (Credit-Bearing Course)..................................... 21
Intermediate Algebra (Credit-Bearing Course) 21
College Algebra (Credit-Bearing Course)... 21
Statistics (Credit-Bearing Course)... 21
Griffin Community College (Semester System) 22
Basic Arithmetic (Developmental Course)... 22
Algebra 1 (Credit-Bearing Course in the Math Department)............. 22
Algebra 2 (Credit-Bearing Course in the Math Department)............. 22
College Algebra with Trigonometry (Credit-Bearing Course in
the Math Department)... 22
Statistics (Credit-Bearing Course in the Math Department) 22
Lester Community College (Quarter System) 22
Fundamentals of Mathematics (Developmental Mathematics).......... 22
Fundamentals of Mathematics I (Credit-Bearing Course in
the Math Department)... 22
Fundamentals of Mathematics II (Credit-Bearing Course in
the Math Department)... 23
College Algebra (Credit-Bearing Course in the Math Department) 23
Statistics (Credit-Bearing Course in the Math Department).............. 23
Instructional Modality for Math Courses 23
Study Guides....What a Concept! 27
Community College Math Student Population in the 1970s 27
The Difference between Lower- and Higher-Level Math Students 28
Learning Disabilities.. 29
Calculator Policy 30
Formal Department Meetings... 30
Relationships with Administration.. 31
Summary.. 32
References 33

3. **The 1980s: Community College Mathematics Reaches
   New Heights** 35
   Participants in This Chapter 35
   Professor Thurmond 35

Professor Bell 35
Professor Johnson 36
Continued Professional Development 36
NADE (National Association for Developmental Education) 36
Course Structure (Circa Mid-1980s) 37
Sisco Community College (Quarter System) 37
    Basic Mathematics (Developmental Math—No Credit) 37
    Algebra 1 (Credit-Bearing Course) 37
    Algebra 2 (Credit-Bearing Course) 38
    Algebra 3 (Credit-Bearing Course) 38
Telford Community College (Quarter System) 38
    Computational Skills (Developmental Math) 38
    Basic Concepts of Algebra (Developmental Math) 38
    Elementary Algebra (Credit-Bearing Course in
    the Math Department) 38
    Intermediate Algebra (Credit-Bearing Course in
    the Math Department) 38
Bordi Community College (Quarter System) 38
    Basic Concepts of Arithmetic (Developmental Math) 38
    Fundamentals of Algebra 1 (Credit-Bearing Course) 39
    Fundamentals of Algebra 2 (Credit-Bearing Course) 39
    Fundamentals of Algebra 3 (Credit-Bearing Course) 39
Habyan Community College (Quarter System) 39
    Basic Mathematics (Developmental Course) 39
    Introduction to Algebra (Developmental Math) 39
    Elementary Algebra (Credit-Bearing Course) 39
    Intermediate Algebra (Credit-Bearing Course) 39
Griffin Community College (Semester System) 40
    Basic Arithmetic (Developmental Course) 40
    Algebra 1 (Credit-Bearing Course in the Math Department) 40
    Algebra 2 (Credit-Bearing Course in the Math Department) 40
Lester Community College (Quarter System) 40
    Fundamentals of Arithmetic (Developmental Mathematics) 40
    Introduction to Mathematics (Developmental Mathematics) 40
    Elementary Algebra (Credit-Bearing Course in
    the Math Department) 40
    Intermediate Algebra (Credit-Bearing Course in
    the Math Department) 40
Changes in Course Structure 41
A Shift to Standardized Placement Testing 42
Improving Student Engagement 43
Preparation Is the Name of the Game 45
A Slightly Changing but Still-the-Same Student Population 46

The Birth of Tutorial Services...47
Emphasis on Study Skills 48
Counseling for Students 49
Method of Instruction 50
Technology for Community College Math Courses in the 1980s 52
Calculators 52
Videos 53
The Emergence of Computer Software...54
Continued Positive Relationships with Administration in the 1980s 54
Summary...55
References 56

4. The 1990s: Mathematics Enters a New Age................................57
Professor McDonald 57
Professor Timlin..57
Professor Mesa 57
Professional Development Grows and Gains More
National Attention..58
    AMATYC (American Mathematical Association of
    Two-Year Colleges)..58
    NADE (National Association for Developmental Education) 58
Course Structure (1995)..59
Basic Concepts of Arithmetic I...59
Basic Concepts of Arithmetic II 59
Rationale for the Restructure 60
General Mathematics...60
Instructional Modalities 61
    Lecture Plus 61
    Phasing Out the Emporium Model..62
    Contextualization...62
    The Development of Distance Learning...63
Continued Efforts to Improve Student Success and Development 67
    The Continuing Development of Math Tutorials 67
    Student Outreach 68
    Women in Transition 69
Calculator Policies of the 1990s...70
The Student Population Ages and Changes 73
The Rise of ESL Students 75
    The Challenges for ESL Students...75
    Teacher Preparatory Classes..76
Shifting Relations between Faculty and Administrators 77
Math Wars Brewing Below...78
Summary...79
References 80

5. The Aughts (2000–2009): A Time of Reform and Turbulence 81
    What Happened? 81
    Returning Participants 81
    Additional Participants 82
        Professor Sutcliffe 82
        Professor Moyer 82
        Professor Trombley 82
        Professor Lopez 82
        Professor Mussina 82
    Developmental Math Gets Bombarded with Statistics 83
    Success Rates Impact State Funding 84
    Persistence Rates Lead to Course Restructure 84
    Lester Community College 85
    Telford Community College 86
    Griffin Community College 86
    Course Restructure Reversed 87
    National Initiatives to Improve Student Success in Math 88
    Acceleration 89
        The Return of the Emporium Model 89
    The Return of Learning Communities 93
    Lecture Becomes a Four-Letter Word 95
    Service-Learning: An Effective but Misplaced Initiative 98
    Supplemental Instruction 100
    Distance Learning Morphs into Online Learning and Explodes 102
    The Challenging Student Population 103
    Higher Demand for Arithmetic Courses 104
    Equity Concerns 105
    Additional Behavioral and Social Issues 106
    Shorter Attention Spans 107
    Student Entitlement 107
    Low Admissions Standards on a National Level 108
    Conflicting Messages from Within 109
    Issues with Uniformity and Too Much Uniformity 110
    Exams 110
    Collaboration 111
    Calculators 112
    Grading 112
    Too Much Uniformity 113
    What about Statistics and College Algebra? 114
    Statistics Moves into the Twenty-First Century 115
    College Algebra Remains Stagnant 116
    Summary 116
    References 117

6. The Teens Part 1: Turbulence and Change Continue............................ 121
New and Returning Participants 121
Professor DeSilva 121
Professor Guzman 121
Professor Holton.. 122
Professor Hickey 122
Professor Williamson 122
Professor Noles.. 122
Professor Douglass 123
The Ambiguity of Intermediate Algebra 123
Complete College America.. 123
The Move from Quarters to Semesters and the Compression
that Followed 124
The Pressure to Compress 124
    SCC.. 125
    BCC.. 127
    TCC.. 128
The Emporium Model Continues... 128
The Decline of Arithmetic Classes 131
The Lowering of Standards in Higher-Level Math Classes?....... 136
Accreditation and Faculty Credentialing Issues 136
Dual Enrollment 138
Multiple Measures 140
Common Core Forms Below 142
Summary.. 143
References 144

7. The Teens Part 2: Alternative Pathways Lead to Signs of Reform 147
New and Returning Participants 147
Professor Harnich... 147
Professor Smith 148
Alternative Math Pathways 148
Statway 148
Quantway... 149
Corequisites 149
The Application of Statway and Quantway................................. 150
    Statway with the Corequisite Has Gone Well........................ 152
    QR with the Corequisite Has Been a Tougher Transition 153
    Group-Based Instruction Aside, QR Has Been Beneficial...... 155
    A Closer Look at Corequisites: The Good, the Bad,
    and the Ugly.. 156
    The Elimination of Standalone Developmental Math 157
    Does Eliminating Standalone Developmental Math Increase
    Completion Rates?... 158
    Arithmetic but No Algebra.. 160

Why the Push to Eliminate? 161
Slim Pickings for Non-Credentialed Faculty.................................... 162
Habyan and Bordi Reverse the Course 162
The Pathways at the End of the Teens 163
Griffin Community College 163
Bordi, Habyan, Telford, and Sisco Community Colleges 164
Assisting ESL Students 164
A Deeper Understanding of ESL Students.. 165
Additional Resources for ESL Students... 166
Professional Development for Math Faculty 167
Continued Challenge 169
Addressing the Equity Gap for Minority Students in Math 169
Where Are the Minority Students Coming From? 169
Outreach to the Inner-City Communities... 170
Assistance and Guidance When Entering School..................................... 172
Educating Faculty 172
General Student Outreach.. 173
Student Profiling... 173
Early Intervention 175
Internet Videos.. 175
Open Educational Resources 176
Reduced Cost or Free Software... 176
Summary... 178
References 178

8. **Teaching during the Pandemic: What We Experienced and
What We Learned** 181
Community College Math Goes Fully Virtual and Online................... 181
The Participants 182
Professor Brown.. 182
Professor Rhodes 182
Professor Stephens.. 182
Conversion to Virtual Instruction 183
Lack of Engagement 184
Volume of Need Too High... 186
Testing 187
What about Arithmetic?... 189
What Worked Well with Virtual Instruction? 190
Higher Comfort Level and Less Anxiety 190
More Time on Task 191
Recorded Sessions... 192
Success Rates for Virtual Instruction.. 193
Explanations 193
The Style and Role of Virtual Classes... 194
Virtual Instruction Can Increase Accessibility for Students................. 195

Instructor Absences 195
Virtual Tutoring 196
Summary 197
Reference 197

9. Learning from the Past and Present 199
Standalone Developmental Math Will Always Be a Part of
Community College Math 199
Full-Length Arithmetic Courses Are Ineffective, but
Addressing Arithmetic Is Essential 200
There Was Culture Change, Which Led to Culture Shock 201
But It Has Always Been and Will Always Be about Money 203
Alternative Math Pathways Have Been the Most Successful
Initiative Thus Far 204
We Take Good Practices and Go Too Far 205
The Emporium Model Will Always Be Around in Some Form 206
Will Online and Virtual Instruction Replace
In-Person Learning? 207
Effective Teaching Involves Well-Rounded Instruction 208
Change Will Never Occur Unless There Is Legitimate
Faculty Buy-In 209
There Is No End in Sight for Student Entitlement and
Student Consumerism 210
While Times Have Changed, Math Content Has Remained
Stagnant 211
Summary 212
References 213

10. Suggested Mathematical Models for Sustaining Success 215
Returning Participants 215
Suggested Pathways for Sustained Success 215
Arithmetic 217
Pathway 1 217
Pathway 2 217
Pathway 3 218
Pre-Algebra/Introduction to Algebra 218
Elementary Algebra 219
Intermediate Algebra 220
Quantitative Reasoning 221
Quantitative Reasoning Booster Course 221
Introduction to Statistics 222
Introduction to Statistics Booster Course 223
College Algebra 224
College Algebra Booster Course 224
Logistics for Booster Courses 225

Length..225
    Virtual Booster Classes?...226
    Instruction...227
Enrollment Flexibility 227
Alternative Math Pathways Reduce Time and Cost for Students.............227
Why Should Students Who Complete Introduction to
Algebra Enroll in a Booster Course?...228
How Much Uniformity Is Necessary for Policies and Assessment?.......228
Average Class Size 229
Additional Initiatives to Sustain Success 230
Supplemental Instruction 231
Learning Communities..231
Utilizing External Resources to Assist Students...........................232
Tutorial Services..232
Academic Advising 232
Student Responsibility 233
Closing Equity Gaps..233
Moving Forward with Math 234
References 235

Appendix A: Design of the Study 237

Appendix B: Lead Questions for Participants (1970s–2009) 241

Appendix C: Lead Questions for the Participants (2000–2019)................243

**Appendix D: Lead Questions for the Participants
(during the Pandemic and Beyond)**......................................245

Appendix E: Demographic Questions...247

Index..249

Preface

A common misconception of higher education faculty is that because we have achieved the ranking of professor, we have always been good and sound students. My experience can debunk that myth very quickly. Throughout most of my elementary and secondary school education, I did not take school very seriously, and I rarely, if ever, put forth maximum effort toward my studies. Consequently, I graduated high school with a "C" average and no goals. However, by my senior year, I knew I needed more education to obtain a better lifestyle, and that is when the open-door policy of community college education welcomed me.

For many students, community colleges are the institutions of second, third, fourth, and fifth chances. All students are accepted and are given the opportunity to succeed. I embraced the diversity of the community college environment. Gone were the cliques from high school. I enjoyed interacting with people of different ages and races.

I knew that I had not been a good student in the past, and I was not a child anymore. This was my last chance to put my best foot forward academically to achieve success in my life. Like other disciplines, math had never been a strength for me. I had no interest in the subject. Quite frankly, I was dreading completing math in community college. Like many students, I placed into developmental math because of my subpar math background. My goal was simply to do what was needed to complete both developmental and college-level math and then move forward. Little did I know, completing math would be a life-changing event.

Not only did I pass basic algebra, but I thrived in the class. I was blessed with an amazing teacher who explained the content so well and helped me develop a love for and an interest in math. With the help of another stellar teacher, I successfully completed my college-level math course as well. By the end of my freshman year, not only had I completed my college math requirement, but I decided on a career path. I wanted to be an educator; moreover, I wanted to teach math at a community college to help students just like me.

I have served as a community college professor for almost 20 years. I enjoy working with students and showing them that they can be successful in math. However, early on in my career, one fact became apparent. As a student, I was one of the fortunate individuals in that I was able to successfully complete my college-level math requirements. This is not the case for many students.

Math has become a barrier for many community college students. This is especially problematic, as all students are expected to complete a math course for their community college associate degree. Moreover, students who intend to transfer to a four-year school must complete a college-level math course that is included in their state's Transfer Module. Such classes include statistics or college algebra.

Throughout the twenty-first century, community college mathematics courses have come under harsh criticism. College algebra, which has historically served as the gatekeeper college-level math course, has a meager 50% success rate and is generally regarded as one of the least successful community college courses (Saxe & Braddy, 2015). However, criticisms have run deeper than college algebra's success rates. The bigger issue has been developmental math. In 2017, the U.S. Department of Education reported that nearly 60% of incoming community college students must complete at least one developmental math course. Wang et al. (2017) added that only 31% of these students complete their required developmental math course sequence. When the developmental math course sequence consists of at least three courses, only 22% of students are successful.

During the 2000s (the aughts) and the 2010s (the teens), researchers contributed to the discipline of community college math suggesting best practices and various ways to improve student success. However, I was attending a meeting, which was led by my college president, Dr. Johnson, and I had a realization. Dr. Johnson stated that for a college to sustain success, that institution must be proactive regarding planning. More specifically, individuals at the institution must plan for effective instructional modality for the future. This led me to ponder: What will make for effective instruction in community college math ten, 20, or even 50 years from now? In summation, success in a discipline, such as math, is not simply about improving present times, it is about being proactive for the future.

In this book, I will be looking at three major entities in the higher education system: developmental math courses, introductory college-level math courses, and the community college. Boylan and Bonham (2007) defined developmental math courses as those below college level. Gatekeeper math courses include developmental math courses, but such courses also include college algebra, quantitative reasoning, and introduction to statistics. More specifically, gatekeeper courses can prohibit students from progressing to college-level classes or block students from completing their math requirements.

For me, a personal critique is that as educators, we do not learn enough from previous experiences or mistakes, and this has been the case in community college math. We continue to design and implement initiatives without much of an attempt to learn from the past. Consequently, in this book, I provide a decade-by-decade analysis of the history of community college math. What did community college math look like? How did we serve students? What struggles did faculty face? What lessons did faculty learn? To provide a rich description, I interviewed faculty who taught community college math in each decade. A total of 30 community college math faculty members shared their experiences from the 1970s through the pandemic of the 2020s. To protect confidentiality, I assigned pseudonyms to all faculty, community colleges, and any people mentioned. In conducting this research and writing this book, I set out to answer the following question: What lessons can we learn from the past and the present to plan for an effective future in community college math instruction?

References

Boylan, H. R. & Bonham, B. S. (2007). 30 years of developmental education: A retrospective. *Journal of Developmental Education, 30*(3), 2–4.

Saxe, K., & Braddy, L. (2015). A Common Vision for Undergraduate Mathematical Science Programs in 2025. Washington, D.C.: Mathematical Association of America. https://www.maa.org/sites/default/files/pdf/CommonVisionFinal.pdf

U.S. Department of Education, Office of Planning, Evaluation and Policy *Developmental Education: Challenges and Strategies for Reform*, Washington, D.C., 2017.

Wang, X., Wang, Y., Wickersham, K., Sun, N., Chan, H. (2017). Math requirement fulfillment and educational success of community college students: A matter of when. *Community College Review, 45*(2), 99–118. https://doi.org/10.1177%2F0091552116682829" https://doi.org/10.1177/0091552116682829

Author

Brian Cafarella, Ph.D., is a mathematics professor at Sinclair Community College in Dayton, Ohio. He has taught a variety of courses ranging from developmental math through pre-calculus. Brian is a past recipient of the Roueche Award for teaching excellence. He is also a past recipient of the Ohio Magazine Award for excellence in education.

Brian has published in several peer-reviewed journals. His articles have focused on implementing best practices in developmental math and various math pathways for community college students. Additionally, Brian was the recipient of the Article of the Year Award for his article, "Acceleration and Compression in Developmental Mathematics: Faculty Viewpoints" in the *Journal of Developmental Education*. Brian also authored the book, *Breaking Barriers: Student Success in Community College Mathematics*.

1

Higher Education: From the Elite to the Mass to the Universal

Mathematics in the American community college system serves a vital role in higher education. Community colleges are accessible to all students, but mathematics is a requirement for all students pursuing a degree. Additionally, gate-keeper mathematics classes such as developmental math classes, college algebra, and introduction to statistics are a challenge to students. Educators must map out effective math pathways for the present and future. However, to accomplish this, we need to understand and learn from the rich past of community college math. This must start with the origin of American higher education.

The Inception of American Higher Education

The inception of American higher education can be traced back to 1636 when Harvard College opened its doors. Harvard was founded by a vote of the Great and General Court of the Massachusetts Bay Colony and was named after its donor the Reverend John Harvard (Thelin, 2004). Harvard followed the European model of higher education. Consequently, most academic books were written in Latin. More specifically, Harvard College expected students to be able to read and interpret Latin poetry and conjugate Greek verbs, as part of a classical education (Butts & Cremin, 1953).

Most colonists attempting higher education were not fluent in Latin and needed assistance (Boylan & White, 1988). Consequently, Harvard College provided tutors, who aided these students (Brubacher & Rudy, 1976). While the term developmental education would not be born for over three centuries, the concept had been established. Higher education has been serving underprepared students since its establishment. However, in these preliminary times, this process was referred to as "tutoring."

The student population in Colonial Times consisted largely of young men intending to join the ministry. Tutors were young men who were more advanced in their studies. Tutors were also lower paid and lower tiered faculty members (Arendale, 2002). Tutoring was informal, as the tutors spent large amounts of the day with their students as they ate with them in dining halls and even slept in the same chamber halls (Brubacher & Rudy, 1976).

DOI: 10.1201/9781003287254-1

It is imperative to note that in Colonial Times, there were no standardized college entrance exams. Higher education served an elite group of young men. Therefore, anyone who had the financial means to attend college could do so.

The Start of Mathematics in Higher Education

Throughout the initial decades of higher education, instruction focused on religious education as well as the classics. However, as the eighteenth century progressed, the need for business skills and mechanical arts increased in certain areas. Consequently, in 1726, Harvard hired its first mathematics professor. Students were taught arithmetic and algebra; however, over time secondary schools began teaching these topics to prepare students for higher education (Willoughby, 1967). In general, college mathematics in the 1700s consisted of Euclidean geometry and algebra. It is noteworthy that at this time secondary schools were sparse; therefore, students entered college as early as 15 (Tucker, 2013). Ultimately, throughout the 1700s, Latin, Greek, and mathematics comprised the classical curriculum. There were still no entrance exams; however, Yale began to require proficiency for incoming students on a standardized arithmetic exam (Arendale, 2002).

By the mid-1800s, the typical college math curriculum was as follows: During freshmen year, students enrolled in algebra and geometry. Students then took more algebra and trigonometry during their sophomore year. Students continuing in the technical fields enrolled in analytical geometry their junior year and possibly calculus their senior year (Tucker, 2013). Additionally, by the 1800s, colleges added the subjects of history, geography, and English to their curricula (Broome, 1903).

The primary method of mathematics instruction, during the seventeenth, eighteenth, and nineteenth centuries, was lecture. The instructor presented the material while students memorized and reiterated such facts on assessments. Again, the American higher education system adapted much of its characteristics from other higher educational systems. The employment of lecture-based instruction can be traced to Christian and Muslim universities during Medieval Times. In general, lecture-based instruction was the primary instructional method worldwide.

The Changing Landscape of Higher Education

In the latter part of the 1800s, the landscape of American higher education changed. During its first two decades, American higher education employed the classical approach to higher education, which consisted of the study of languages, philosophy, and eventually mathematics. This was based on the European model of education, which sought to train the mind. However, by the mid-1800s, students began to lose interest in this model. Consequently, in 1862, the Morrill Land Grant Act was established. This act, passed by U.S. Congress, provided grants of land to states to finance the foundation of colleges that specialized in agriculture and mechanical arts. This opened the doors of higher education to farmers and other working-class individuals who had been excluded from higher education. This act was significant as states began to fund higher education. In 1890, the Second Morrill Land Grant Act was established. This act provided funding for 19 public Black colleges (Allen & Jewell, 2002). Additionally, in the late 1800s, the first generation of women entered higher education enrolling primarily in nursing, education, and social work programs (Thelin, 2004).

Throughout its first two centuries, American higher education served elite males who could afford a college education. However, by the late 1800s, the American higher education system was serving those who had previously been omitted: students of color and women. This led to a massive increase in attendance in higher education. By 1900, enrollment at America's colleges and universities more than doubled from 1870 (Arendale, 2002).

During the seventeenth and eighteenth centuries, tutors assisted the under-prepared college students. However, by the mid-1800, the traditional tutors were in shorter supply. Young men were utilizing higher education less for entry into the ministry and more for pathways to employment in technical and agricultural fields. Additionally, the student population in higher education continued to grow and serve a more diverse population (Arendale, 2002). Consequently, institutions sought new ways to serve underprepared students. In 1849, the University of Wisconsin founded America's inaugural remedial or college-preparatory program. This program offered courses in reading, writing, and arithmetic for underprepared college students (Brier, 1986). Other institutions soon followed suit.

As higher education's population continued to grow, so did the number of underprepared students. By the late 1800s, more than 80% of America's colleges and universities offered a college preparatory program (Brier, 1986). Students entered higher education as young as 14 years of age. Consequently, many students entered higher education with very deficient reading and mathematics skills. Many students required remedial education, as they could not read, write, or perform basic arithmetic (Maxwell, 1997). Why did colleges have such low standards? Land Grant institutions opened rapidly, and this resulted in intense competition for enrollment, as these institutions needed to address startup costs. Consequently, these colleges were not concerned with incoming academic ability for students (Arendale, 2002).

The need for remedial education often lengthened students' time to complete baccalaureate degrees to 6 years or longer (Cassaza & Silverman, 1996). By 1894, more than 40% of incoming college and university students were enrolled in a remedial course (Ignash, 1997). As the number of remedial classes increased, the demand for faculty also increased. In general, university professors resented and looked down on remedial instruction (Arendale, 2002).

The need to educate minority students increased the need for remedial education. Again, the second Morrill Land Grant Act of 1890 opened the doors for many minority students. However, the American Missionary Society also established various colleges to educate freed slaves (Brubacher & Rudy, 1976). With few secondary schools available and many schools closed to freed slaves, remedial education in higher education was their only option.

To increase college preparedness and reduce the need for remediation, Columbia University established the College Entrance Examination Board (CEEB). The CEEB organized standardized exams, such as the SATs, for secondary schools to ensure college readiness. Students pursued mathematics in college to obtain positions in the engineering and actuary fields but were underprepared. Therefore, despite the establishment of the CEEB, there was an increase in the need for remedial arithmetic and algebra at the college level. It is also noteworthy that at this point, colleges were still tuition-driven and continued to offer more remedial courses to accommodate students and increase enrollment.

The Progressive Movement

As the nineteenth century progressed, American higher education moved away from the classical curriculum and toward a general education curriculum. Enrollment in higher education continued to decrease, as students did not see the value in a classical education. This included mathematics, as students were bored with the drill and practice of the discipline. Consequently, mathematics had become optional in secondary schools, and by the turn of the twentieth century, math courses were not required in many colleges and universities. The only students who enrolled in mathematics were those intending to become engineers, actuaries, or mathematics teachers (Grouws, 1992).

The decreased emphasis on mathematics was due to the progressive movement, led by educational specialists such as John Dewey, of the late nineteenth and early twentieth century. Progressives stressed the belief that schools should focus more on socialization as well as critical thinking and problem-solving skills (Tucker, 2013). As opposed to lecture-based instruction and drill and practice, progressives emphasized group-work and cooperative learning. Additionally, progressives stressed that learning should be personalized for each student. Many progressives did not view the drill and practice of math as relevant to the real world (Dewey, 2008).

Not everyone agreed with the gravitation away from mathematics requirements. In 1915, the Mathematical American Association (MAA) was established with the purpose of focusing on teaching practices and standards for mathematical instruction. Additionally, educational activists such as E.B. Wilson called for all college students to complete required mathematics courses.

The Two-Year Junior College

In the late 1800s, several university presidents led by William Rainey Harper, president of the University of Chicago, pushed for secondary schools to offer general education courses that students could complete during their freshmen and sophomore years of college. Ultimately, these university presidents did not even consider such courses to be at the university level. Therefore, Harper proposed that universities should add extensions, which would include a 13th and 14th year. In such extensions, high school graduates could complete these general education courses. When students completed these classes, they could then apply to the university to enroll in more applied and degree-focused classes. Ultimately, this group of college presidents asserted that this model would lead to bachelor's degrees that were more research-focused and scholarly (Drury, 2003; Scherer & Anson, 2014).

While some university presidents supported Harper's proposal, it did not come into fruition. Several college presidents expressed concern that these university extensions, labeled as junior colleges, would set a bad precedent. Their concern was that universities would lessen the rigors of their admissions requirements for students in junior colleges (Scherer & Anson, 2014). Ultimately, this proposal went against the mindset that college was for the elite.

Harper's original proposal for junior colleges failed; however, the concept of the junior college succeeded. Harper's friend, Joseph Stanley Brown, the superintendent of Joliet Township High School, created an experimental post-graduate high school program. This led to the establishment of Joliet Community College in 1901, which ultimately led to the standalone junior college (Griffith, 1976; Henderson, 1960).

In the early twentieth century, junior colleges evolved into standalone institutions. Students who entered junior college spent the first 2 years completing remedial and general education classes (Eells, 1931). Junior colleges served various purposes. Ultimately, these institutions prepared students for university baccalaureate degree programs while allowing them to complete their general education requirements. However, junior colleges also offered education for students in sparsely populated areas, especially rural and farming communities. Some junior colleges were also affiliated with religious denominations (Drury, 2003).

Junior colleges reached their zenith in the 1940s. By 1940, 11% of all college students were enrolled in a junior college (Geiger, 2005). Nationally, there were about 350 private junior colleges that served over 100,000 students (Weisbrod et al., 2008).

A Focus on Math and a Greater Need for Developmental Math

The study of mathematics was deemphasized during the first part of the twenty-first century; however, the tides began to turn during the Great Depression. The Great Depression negatively impacted higher education. Colleges faced low enrollment, and some were even forced to close. However, enrollment in mathematics courses gradually increased. Recognizing the importance of mathematics, and to increase enrollment, some colleges offered accelerated programs for career training that utilized mathematics. This served as a precursor to the great mathematics surge post World War II.

The period following World War II greatly impacted mathematics in higher education, the need for developmental education, and higher education in general. Following the conclusion of the war, the U.S. government was fretful about the potential unemployment of millions of veterans. Consequently, the government implemented the Serviceman's Readjustment Act, also known as the G.I. Bill of Rights. This act allotted millions of dollars to allow veterans to attend college and further their education, and subsequently, millions of veterans took advantage of this act (Olson, 1974; Thelin, 2004). While institutions of higher education were able to be more selective by the mid-twentieth century, many allowed the World War II veterans to enter. However, colleges needed to accommodate these students by offering additional remedial classes to meet their needs. After all, these students had been out of school for some time and needed refresher classes, especially in arithmetic and algebra.

The nation's interest in math continued to grow post-World War II. In 1953, the MAA formed the Committee on the Undergraduate Program in Mathematics with an initial goal of standardizing freshmen-level mathematics. Consequently, this committee developed a course entitled, "universal mathematics," which consisted of concepts such as mathematics of sets, limits and derivatives, and counting and probability. Additionally, in 1957, the Soviet Union launched Sputnik. This started the race of the space age between the United States and the U.S.S.R. Consequently, throughout the late 1950s and 1960s, the number of students who were interested in studying mathematics and engineering increased greatly (Tucker, 2013). In summation, America viewed learning mathematics as a common good. World War II and the launching of Sputnik allowed America to see the relevance of mathematics to the modern world (Tucker, 2013). As opposed to training the mind, as it was in 1700s, excellence in mathematics became a means toward national pride but also national security (Barlage, 1982).

In 1958, America passed the National Defense Education Act. This act, in part, allotted money for mathematics education. Consequently, the National Science Foundation established the School Mathematics Study Group. One goal of this group was to allot updated and better-quality textbooks for mathematics students (Barlage, 1982). Student-interest in mathematics continued to surge during the 1960s. In fact, the number of mathematics bachelor's degrees doubled in this time. By 1970, 4% of students who achieved bachelor's degrees were mathematics majors (National Science Foundation, 2008).

The increased student interest in mathematics and the push for all students to complete calculus put more focus on college algebra and statistics. While college algebra content varied between institutions, the content in this course (e.g., various functions, trigonometry, conic sections) was imperative for preparing students for calculus and more abstract mathematics. Subsequently, enrollment in college algebra soared in higher education (Taylor, 2017). Additionally, statistics were used widely throughout the war, and in general the nation garnered a greater appreciation for the use of statistics. Consequently, the number of collegiate-level statistics classes increased (Tucker, 2013).

The Rise of the Community College

In its inaugural time, higher education served an elite population of young men. However, through such events as the two Morrill Land Grant Acts, the entry of women in higher education, and the establishment of the G.I. Bill, by the mid-twentieth century, higher education served the mass. However, by the late 1960s, American higher education became accessible to the universal student population.

In the 1960s, there were various civil rights movements in America. Such movements attempted to close the gaps of equality and more specifically serve those who had been marginalized. This was the case in higher education as well. In 1965, President Lyndon Johnson signed the Higher Education Act (HEA). The primary goals of the HEA were to provide funding for small and less developed colleges and to increase access to higher education, primarily for lower class and minority students. In the mid-1960s, the federal government established various student-loan programs. In the 1970s, this was followed by Pell grants for students in need of academic assistance. The HEA also sought to help non-traditional students achieve a college degree, as in 1964, less than 10% of people 25 and older earned a college degree (Cohen & Brawer, 2008). In summation, the HEA provided additional pathways to higher education for students from lower socioeconomic backgrounds as well as non-traditional students.

As higher education looked to provide accessibility to more students, America saw the growth of the public community college. As mentioned earlier, 2-year junior colleges had been in existence for most of the twentieth century. Like junior colleges, the goal of public community colleges was to prepare students for 4-year universities and to provide education for students in sparsely populated areas. However, public community colleges provided open access for students with low tuition. In summation, community colleges served the universal population.

Community colleges became a fixture in higher education very quickly. In the 1960s, community colleges grew more rapidly than any institution of higher education (Drury, 2003). From 1965 to 1972, community colleges opened at a rate of one per week (Geiger, 2005). In the 1970s, enrollment at community colleges increased from 1.6 million to 4.5 million (Brint, 1989). Additionally, the explosion of the open access public community college contributed to the decline of the private 2-year junior college. By the late 1980s, nearly 75% of junior colleges disappeared, and less than 1% of 2-year college students were enrolled in these institutions.

In their inaugural years, community colleges received their primary funding from their state governments, and such funding was based on student full-time equivalent (FTE). An FTE is calculated by taking the total number of credits carried by all students and dividing by the number of credits in a full-time load (this is generally 12 or 15). The greater the FTEs, the more funding a community college received.

Since their rise in the 1960s, community colleges have advertised themselves as open admissions institutions. However, during the latter part of the twentieth century, various community colleges had some admissions requirements. Some community colleges restricted admissions for students without a high school or equivalency diploma. Other schools did not require such degrees but mandated that enrolling students be at least 18 years old. In summation, there has never been a standard definition for an open admissions or open-door policy (Scherer & Anson, 2014).

Like junior colleges, community colleges offered remedial math courses as well as math courses equivalent to freshmen and sophomore level. Regarding the structure of course content, states began founding committees known as the board of regents. These were committees that established guidelines for college courses and standards regarding content covered in these courses. For example, the Ohio Board of Regents was founded in 1963; the Tennessee Board of Regents was established in 1972; the Alabama Board of Regents was founded in 1969; the Mississippi Board of Regents was established in 1944. Such guidelines helped colleges establish content for courses such as college algebra, elementary statistics, and remedial courses such as elementary and intermediate algebra.

The Uniqueness of the Community College Model

The American community college system is unique to higher education worldwide. Community colleges have historically offered low tuition for students. During the 1963–1964 academic year, students paid an average of $34 per credit hour at American community colleges. By, the 2020–2021 academic year, this cost had risen to only $142 per credit hour. This amount pales in comparison to the $395 per credit hour at public 4-year schools and greatly contrasts to the $1,368 per credit hour at private universities (Hanson, 2021). However, its low cost does not make the American community college unique. As of 2021, in countries such as Argentina, Brazil, Denmark, Germany, Iceland, Norway, Sweden, and the United Kingdom, higher education is free to citizens. However, schools in these countries contain admissions requirements such as prior grade point average, successful completion of an entrance exam, or at least the possession of a secondary education degree. The American community college accepts all students regardless of their academic background. Additionally, the American community college system is unique in that students can study a variety of academic disciplines such as mathematics, English, engineering, nursing, paralegal studies, or social work. However, students may also complete various certificates in corrections, hospitality, coaching or vocational studies.

Summary

The American Higher Education system was inaugurated in 1636. At that time, higher education served an elite student population who could afford an education. Moreover, higher education began as vehicle for those striving to join the ministry. However, by the latter part of the twentieth century, higher education served the universal population. This was a slow paradigm shift as various anomalies went against the original mindset of the American higher education model.

Throughout the eighteenth, nineteenth, and twentieth centuries, higher education evolved. As Land Grant institutions and junior colleges opened, higher education began serving a larger and more diverse student population. Higher education's emphasis on mathematics fluctuated as well. In the 1700s, mathematics was part of a classical curriculum aimed to train the mind. As the emphasis on the classical curriculum waned and progressives influenced education, mathematics was deemphasized and required less in higher education. However, World War II and the launching of Sputnik brought a new appreciation to mathematics and led to the nation viewing math as an imperative discipline.

Since its inception, serving the underprepared student has been a constant issue, as students have entered higher education deficient in their academic skills and have required some sort of remedial education. Attempts by higher education to eliminate remedial education have failed, and the number of remedial courses increased.

To make education accessible to the universal student population, the open access community was established in the 1960s. Community colleges were available to students at a low cost and more importantly were ready to serve the previously marginalized student. With that the American community college and the discipline of mathematics were about to be connected and serve a widespread student population.

References

Allen, W. R., & Jewell, J. O. (2002). A backward glance forward: Past, present, and future perspectives on historically black colleges and universities. *The Review of Higher Education, 25*, 241–261. http://dx.doi.org/10.1353/rhe.2002.0007

Arendale, D. (2002). History of supplemental instruction (SI): Mainstreaming of developmental Education. *Center for Research on Developmental Education and Urban Literacy,* General College, University of Minnesota.

Barlage, E. (1982). The new math: A historical account of the reform of mathematics in the United States of America. (ERIC Document Reproduction Services No. ED 224703).

Boylan, H. R., & White, W. G. (1988). Educating all the nation's people: The historical roots of developmental education part I. *Research in Developmental Education, 4*(4), 14. National Center for Research on Teacher Learning.

Brint, S., & Karabel, J. (1989) *The Diverted Dream: Community Colleges and the Promise of Educational Opportunity.* Paris: Oxford University Press.

Broome, E. C. (1903). *Historical and Critical Discussion of College Admission Requirements.* MacMillan.

Brier, E. (1986). Bridging the academic preparation gap: An historical overview. *Journal of Developmental Education, 8*(1), 2–5.

Brubacher, J. S. & Rudy, W. (1976). *Higher Education in Transition.* Harper & Row: New York.

Butts, R. F., & Cremin, L. A. (1953). *A History of Education in American Culture.* Henry Holt & Company.

Cassaza, M. E. & Silverman, S. E. (1996). *Learning Assistance and Developmental Education: A Guide for Effective Practice.* Josey-Bass.

Cohen, A. M. & Brawer, F. B. (2008). *The American Community College* (5th ed). Jossey-Bass.

Dewey, J. (2008). *Democracy and Education.* Filiquarian Publishing.

Drury, R. L. (2003). Community colleges in America: A historical perspective. *Inquiry, 8*(1), 1–6.

Eells, W. C. (1931). *The Junior College.* Houghton Mifflin.

Geiger, R. L. (2005). The ten generations of American higher education. In P. G. Altbach, R. O. Berdahl, & P. J. Gumport (Eds.), *American Higher Education in the 220 Twenty-First Century: Social, Political, and Economic Challenges* (pp. 36–70). The Johns Hopkins University Press.

Griffith, W. S. (1976). Harper's legacy to the public junior college. *Community College Frontiers, 4*(3), 14–20.

Grouws, D. A. (Ed.). (1992). *Handbook of Research on Mathematics Teaching and Learning: A Project of the National Council of Teachers of Mathematics.* Macmillan Publishing Co, Inc.

Hanson, M. (2021, September 4). *Cost of a college class or credit hour.* Education Data.org. https://educationdata.org/cost-of-a-college-class-or-credit-hour

Henderson, A. D. (1960). *Policies and Practices in Higher Education.* Harper and Row.

Ignash, J. M. (1997). Who should provide postsecondary remedial/developmental education?. *New Directions for Community Colleges, 100,* 5–20.

Maxwell. M. (1997). *Improving Student Learning Skills* (Rev. ed.). H & H Publishing.

National Science Foundation (2008). *S & E Degrees: 1966–2006* (Report 08–31).

Olson, K. W. (1974). *The G.I. Bill, The Veterans, and The Colleges.* University Press of Kentucky.

Scherer, J. L., & Anson, M. L. (2014). *Community Colleges and the Access Effect: Why Open Admissions Suppresses Achievement.* Palgrave Macmillan.

Taylor, S. (2017). *The evolution of College Algebra: Competencies and Themes of a Quantitative Reasoning Course at the University of Kentucky [Doctoral dissertation, Western Kentucky University].*

Thelin, J. R. (2004). A History of American Higher Education. Johns Hopkins University Press.

Tucker, A. (2013). The history of the undergraduate program in mathematics in the United States. *The American Mathematical Monthly,* 1–21. doi: 10.4169/amer.math.monthly.120.08.689.

Weisbrod, B. A., Ballou, J. P., & Asch, E. D. (2008). *Mission and Money: Understanding the University.* Cambridge: Cambridge University Press.

Willoughby, S. (1967). *Contemporary Teaching of Secondary School Mathematics.* John Wiley & Sons.

Hansen, D. (2021). Reimagining the ethical dimensions of equitable education: Past and future. Educational Studies, 57(4), 411–426.

Heidegger, A. H. (2021). Time and human beings in school settings [Paper presented]. Rethinking education. Why *the* school: The good knowledge science summit. *Journal of educational studies*.

Marzano, R. (2007). The art and science of teaching. ASCD Books.

Noyes, I. (Ed.). (2016). *How to change the world*. Phaidon Press.

Oreos, R. (2014). Spoken from faith. The classroom as an open space. Denver University Press.

Sheehan, L., & Anson, M. (2011). Community college ideas. Teachers and support in longitudinal perspective. Palgrave Macmillan.

Taylor, S. (2017). The failure of college education. *Engagement and retention*. Cambridge Scholars Publishing. Oxford University Press.

Thelin, J. R. (2019). *A history of American higher education*. Johns Hopkins University Press.

Tobin, A. (2016). The history of the undergraduate classroom. Contribution to the United States. *American educational history*, 12(1), 210–229.

Wadhwa, R. A. (Ed.), & Reynolds, D. (2000). *Common standards of cumulative education*. Cambridge University Press.

Willingham, D. (2021). Cognition as a function of the theory and the mind. *Atlas studies*. John Wiley & Son.

2

The 1970s: Mathematics and the Community College Become Acquainted

By the 1970s, community colleges were growing and thriving. Community college math courses were now open to the universal student population. This included students who needed math courses for their future careers or students who simply needed math for their degree requirements. Whether well-prepared, underprepared, or severely underprepared, community colleges were ready to meet the math needs of these students. In this chapter, I will be exploring community college math in the 1970s. How was the content delivered? What did instructional modality look like? Was the student population very different from the twenty-first century? Overall, how did the climate of community college math look in this period compared to the twenty-first century?

The Participants

To obtain a better idea of the ambiance of community college math during the 1970s, I interviewed seven faculty members who instructed gatekeeper community college math courses (developmental or introductory college-level) during this period.

Professor Ballard

Professor Ballard instructed developmental and college-level math from 1973 to 2009 at Sisco Community College. He has a bachelor's degree and a master's degree in mathematics. Professor Ballard spent his entire career as a tenured professor. At Sisco Community College, the developmental math and college-level math have always been housed under one department. At the time Professor Ballard started, he relayed that his average math class size hovered around 30 or 35 students.

DOI: 10.1201/9781003287254-2

Professor Mitchell

Professor Mitchell taught both developmental math and college-level math at Telford Community College from 1977 to 2014. She taught full-time from 1977 to 2008 and as an adjunct through 2014. At Telford Community College, developmental education, which contained developmental math, was a separate department from the college-level math department. Professor Mitchell served as coordinator for developmental math from 1990 to 2006, as developmental math was her home department. Professor Mitchell has a bachelor's and a master's degree in math along with a Ph.D. in education. When Professor Mitchell started, her average class size was around 70.

Professor Milacki

Professor Milacki instructed both developmental and college-level math full-time at Bordi Community College from 1971 to 2008. At Bordi Community College, developmental math and math are housed under one department. Professor Milacki served as chairperson of the math department from 1989 to 2006. Professor Milacki has a bachelor's and a master's degree in math. Professor Milacki stated that when she started, the average class size for most math classes consisted of 70– 80 students.

Professor Morgan

Professor Morgan instructed both developmental and college-level math, full-time, from 1977 to 2011 at Habyan Community College. At Habyan Community College, developmental math and college-level math are contained under the same department. Professor Morgan served as chair of the math department from 1993 to 2005. She has a bachelor's and a master's degree in math. When Professor Morgan began teaching, the average class size for most classes was 50–60 students.

Professor DeLeon

Professor DeLeon started teaching developmental math at Griffin Community College in 1975. She also served as chair of developmental education from 1990 to 2006. She retired in 2008. Professor DeLeon has a bachelor's degree in math education and a master's degree in education. The developmental math and math department were in separate departments. At Griffin Community College, when Professor DeLeon started, the average class size was 35–40 students.

Professor Fenimore

Professor Fenimore taught primarily developmental math at Lester Community College from 1977 to 2007. She taught as an adjunct from 1977 to 1982 before being hired as a full-time professor. Professor Fenimore served as coordinator of developmental math from 2000 to 2006. At Lester Community College, developmental and college-level math were separate departments. Professor Fenimore has a bachelor's and master's degree in math education. When Professor Fenimore started at Lester, the developmental math average class size was 30 students.

Professor Wallace

Professor Wallace taught college-level math, full-time, at Lester Community College from 1972 to 2007. He served as department chair from 1990 to 2007. Professor Wallace's bachelor's and master's degrees are in mathematics. When Professor Wallace began teaching, the average class size in college-level math courses ranged from 25 to 30 students.

Professional Development

A consistent theme throughout the 1970s and beyond was the growth of professional development at the community college level. The participants in this study asserted that pedagogical practices and the instructional modalities in community college math developed immensely in this time due to networking with other colleagues and studying innovative practices.

Several participants discussed the impact of the American Mathematical Association of Two-Year Colleges (AMATYC). In 1967, the Mathematics Association of Two-Year Colleges Journal was started by George Miller and Frank Avenoso. This journal described the need for and encouraged the collaboration of mathematical organizations focusing on pedagogical practices for two-year institutions. Consequently, various states such as New York, Pennsylvania, and Oklahoma began forming their own organizations that focused on the improvement of instruction and effective communication among instructors and institutions (Blair, 1999).

In 1974, John Massey, from Chesapeake Community College in Maryland, led a committee to start AMATYC, an organization that focused on the development and improvement of mathematics at two-year colleges across America. In 1975, AMATYC held its first annual meeting in Chicago with approximately 300 people. As developmental math was a major part of math programs at community colleges, AMATYC took an active approach to developmental math as well. Consequently, AMATYC president, Herb Gross, established the Developmental Mathematics Curriculum Committee (Blair, 1999).

Placement Testing

Throughout the history of higher education, colleges and universities utilized various entrance exams to assess the skill level of incoming students. However, it was not until the 1980s that states began mandating placement exams and colleges began employing computerized exams. How did community colleges handle placement testing for incoming students during their inaugural years? In general, community colleges developed locally utilized assessments. I asked the participants about their recollections regarding the early days of community college placement testing.

> I was on the committee that developed our math placement test at Bordi [Community College], *said Professor Milacki*. I still remember it; it was a 100-question test that covered questions from arithmetic to calculus. After the students took the test, they would wait while someone in the testing center graded it by hand and then informed the students how they did and what class they needed to register for.

Professor Morgan also served on the committee that designed the math placement exams at Habyan Community College.

> We didn't have to redesign it [the placement test] too often, maybe every two or three years. We basically went by our final exams. Most of the placement test questions were similar to the questions on the final exams for courses like basic algebra and college algebra. It was pretty simple. If students couldn't get the questions correct on the placement test; they weren't ready for that course.

Professor Milacki elaborated.

> Students got a raw score based on how many problems they got correct. We had specific cut scores to get into each class, kind of like now.

In other cases, placement tests were standardized among cohorts of community colleges.

Professor Mitchell expounded:

> I remember once in the late [19]70's there were about two or three of us from each community college who got together to design a placement test that all of us could use. I believe there were five total community colleges that used the same placement test. I suppose the goal was to have some standardization to make sure we were all doing the same thing. We all traveled to one school, and we sat in a room till we came up with a standardized 120-question test.

I asked Professors Morgan, Milacki, and Mitchell if they felt their students were academically prepared for the math course in which they started.

> I had class sizes of 50 or 60 students at the time, and it was hard to assess the skill-level when they came in. I had a lot of students drop out after a few weeks, but I'm not sure if they were unprepared for the class or were just unmotivated. Maybe both, explained Professor Morgan.
>
> I think most were [prepared], said Professor Milacki. I liked that we had open ended questions on our placement tests, no multiple choice. I liked that our placement test questions came from our courses.
>
> To be honest, I don't really remember if they were unprepared. I had 70 or 80 students in a class, and I was really overwhelmed. I was 25 years old; it was my first job, and I was trying to keep my head above water. What I do remember is more than half the class dropped out by the end of the quarter, so I had a lot of students not doing what they were supposed to do, admitted Professor Mitchell.

Throughout the twentieth century, most colleges and universities utilized some sort of incoming placement exam. However, during the 1960s, some institutions began using a laissez-faire approach of omitting placement exams and allowing students to start at the math class of their choosing (Cohen & Brawer, 2008). It is noteworthy that up to the 1960s, many colleges followed the concept of in loco parentis, which means that colleges acted in place of the parent where they saw fit. However, in the 1960s, there were widespread student revolutions to express their desire for freedom, which consequently led to increased student freedom (Thelin, 2004).

Professor DeLeon recalled this being the case when she started teaching at Griffin Community College.

> I don't know why but we had no placement test for students starting at Griffin. Students would meet with their advisors and discuss the [math] class they should start in. Then, on the first day of class, we would give them a diagnostic test for that class. I think it was about fifteen questions. If the student scored too low, we would have them moved to a lower course and if they scored really high, we would give them the option to move into a higher math class. I know it sounds like a lot of work with all the technology we have now, but we made it work.

Professor DeLeon commented on the readiness level of the students.

> I guess eventually it was OK. It was just chaotic for the first couple of
> weeks. People were switching classes like crazy. I remember this one
> girl started in intermediate algebra, then went to basic algebra and then
> to basic math because each class was too hard. We also had a lot of
> attrition as well. I'd say at least half the class was gone by the middle
> of the semester.

Sisco Community College had this laissez-faire policy in place when Professor
Ballard started teaching there in 1973.

> I think students met with their counselors to discuss what class they should
> start in, but I'm not sure. No, we didn't give them any kind of first-day diag-
> nostic. They showed up on the first day and we tried to teach them.

Professor Ballard elaborated on the preparedness level of his students.

> It was a mess. I had students in college algebra who didn't even know
> their signed numbers or how to combine like terms or how to evaluate
> expressions. I had students in my introduction to statistics classes who
> could not do operations with fractions. I had some students who could
> barely read.

I asked Professor Ballard if he thought this was unusual.

> I guess it was a little strange. I went to a university, and I never saw any-
> thing like that, but I was also 24 years old and completely green. I had no
> idea what to expect when I started teaching, so I thought it was normal.
> After a few weeks the lower [level]students gradually dropped out. I'd say
> on average about 40% of the students passed those classes.

By the late 1970s, both Griffin and Sisco Community College abandoned the
laissez-faire approach for placement testing. Professor DeLeon elaborated

> It was actually the chair of the math department who came up with the
> idea of having a school placement test. I still remember him presenting
> the idea to the math and the developmental math department at a meet-
> ing. He got the idea from going to AMATYC. I remember because I had
> never heard of AMATYC before that meeting. I wasn't on the committee,
> but I believe there were faculty from developmental math and the math
> department who worked on the placement test.

Professor Ballard shared:

> I was relieved when I heard we were implementing a placement test.
> I had never thought about it, but it made sense. It did make a differ-
> ence. I mean, we still had people who were unprepared, but at least I had
> people in my college algebra class who could add fractions and people in
> my statistics class who could read.

Course Structure

What did the course sequencing look like for developmental math and introductory math courses in the 1970s? Below is the content for such courses in the mid-1970s. It is noteworthy that classes on the quarter system lasted 10 or 11 weeks, whereas semester-length classes were and are 15–16 weeks.

Sisco Community College (Quarter System)

Basic Mathematics (Developmental Math—No Credit)

This class focused on all arithmetic operations (whole numbers, fractions, decimals, percentages), rational numbers, simplifying algebraic expressions, solving simple and complex algebraic equations, laws of exponents, and factoring.

Algebra 1 (Credit-Bearing Course)

Algebra 1 consisted of operations with rational numbers, polynomials, laws of exponents, complex algebraic equations, rational expressions, inequalities, graphs of linear equations, and roots and radicals.

Algebra 2 (Credit-Bearing Course)

This course emphasized system of linear equations, quadratic equations, complex numbers, parabolas, and absolute value equations

College Algebra (Credit-Bearing Course)

The content in this course focused on quadratic equations, functions, roots and radicals, inequalities, logarithms, binomial theorem, conic sections.

Statistics (Credit-Bearing Course)

This course focused on graphs and tabular presentations of data, probability, and statistical distributions, sampling, permutations, and combinations, and applications of statistical analysis.

Telford Community College (Quarter System)

Fundamentals of Mathematics (Developmental Math)

Fundamentals of Mathematics emphasized operations with whole numbers, fractions, decimals, ratio and proportions, percentages, signed numbers, linear equations, operations with polynomials.

Elementary Algebra (Credit-Bearing Course in the Math Department)

This class focused on linear equations, operations with polynomials, laws of exponents, factoring, rational expressions, roots and radicals, and graphs.

Intermediate Algebra (Credit-Bearing Course in the Math Department)

This course emphasized system of linear equations, quadratic equations, parabolas, complex numbers, roots and radicals.

College Algebra (Credit-Bearing Course in the Math Department)

College algebra consisted of quadratic equations, operations with functions, permutations and combinations, logarithms, sequences and series, and conic sections.

Statistics (Credit-Bearing Course in the Math Department)

This class focused on data description, probability, hypothesis testing, confidence intervals, and sampling.

Bordi Community College (Quarter System)

Basic Concepts of Mathematics (Developmental Math)

This course focused on operations with whole numbers, fractions, decimals, ratios and proportions, units of measurements, operations with rational numbers, evaluating expressions, solving linear equations, and problem-solving applications.

Fundamentals of Algebra 1 (Credit-Bearing Course)

This class emphasized operations with rational numbers, solving linear equations, solving quadratic equations, laws of exponents, factoring, and rational expressions.

Fundamentals of Algebra 2 (Credit-Bearing Course)

Emphasized in this class were simultaneous equations [system of equations], roots and radicals, absolute value equations, complex numbers, and problem-solving applications.

College Algebra (Credit-Bearing Course)

This class focused on operations with functions, permutations and combinations, the binomial theorem, operations with matrices, logarithms, sequences and series.

Statistics (Credit-Bearing Course)

Statistics focused on graphical and tabular presentations of data, set theory, probability, parameters, confidence intervals, hypothesis testing, and statistical distributions.

Habyan Community College (Quarter System)

Basic Mathematics (Credit-Bearing Course)

This course highlighted operations with arithmetic (whole numbers, fractions, decimals, percentages, proportions), signed numbers, evaluating expressions, linear equations, laws of exponents, and factoring.

Elementary Algebra (Credit-Bearing Course)

This class focused on solving equations, literal expressions, laws of exponents, factoring, rational expressions, and operations with polynomials.

Intermediate Algebra (Credit-Bearing Course)

Highlighted in the course were quadratic equations, simultaneous equations, complex numbers, roots and radicals, inequalities, relations and functions.

College Algebra (Credit-Bearing Course)

This course emphasized relations and functions, operations with matrices, logarithms, sequences and series, and conic sections.

Statistics (Credit-Bearing Course)

This class focused on data description, set theory, probability, parameters, chi-square test, permutations and combinations, and hypothesis testing.

Griffin Community College (Semester System)

Basic Arithmetic (Developmental Course)

This course emphasized operations with whole numbers, fractions, decimals, proportions, units of measurements, percentages, and operations with integers.

Algebra 1 (Credit-Bearing Course in the Math Department)

This class highlighted operations with integers and rational numbers, evaluating expressions, linear equations, system of linear equations, laws of exponents, factoring, and operations with rational expressions.

Algebra 2 (Credit-Bearing Course in the Math Department)

Emphasized in the class are operations with polynomials, quadratic equations, roots and radical equations, complex numbers, inequalities, absolute value equations, relations and functions, and applications.

College Algebra with Trigonometry (Credit-Bearing Course in the Math Department)

This course focused on functions, logarithms, trigonometric functions, laws of sines and cosines, trigonometric identities, sequences and series, conic sections.

Statistics (Credit-Bearing Course in the Math Department)

Emphasized in this class were set theory, probability, data description, chi-square test, sampling, statistical distributions, confidence intervals, and hypothesis testing.

Lester Community College (Quarter System)

Fundamentals of Mathematics (Developmental Mathematics)

This class focused on all arithmetic skills, percentages, ratios and proportions, equations, laws of exponents, radicals, and quadratic equations.

Fundamentals of Mathematics I (Credit-Bearing Course in the Math Department)

This course emphasized operations with signed numbers, laws of exponents, literal expressions, monomial and polynomial expressions, linear equations, factoring, and algebraic fractions.

Fundamentals of Mathematics II (Credit-Bearing Course in the Math Department)

Highlighted in this class were linear and quadratic equations, exponents, radicals, ratios and proportions, and logarithms.

College Algebra (Credit-Bearing Course in the Math Department)

This class focused on linear and quadratic equations, system of equations, exponents, radicals, complex numbers, higher degree equations, inequalities, progressions, permutations and combinations, probability, binomial theorem, and determinants.

Statistics (Credit-Bearing Course in the Math Department)

This course focused on an introduction to statistical techniques and methodology, graphical and tabular presentation of data, parameters, statical distributions, sampling, confidence limits, and tests of hypothesis.

Instructional Modality for Math Courses

I asked the participants about the instructional modality as well as the delivery methods for community college gatekeeper math courses in the 1970s. Many classes followed the traditional face-to-face and lecture method that had been in American higher education throughout the centuries.

> All our classes were face-to-face, and it was straight lecture. Students came to class, and I lectured. I didn't know any other way; it was how I learned math, so I figured that's how I was supposed to teach, reflected Professor Ballard.
>
> It was exclusively face-to-face. It was lecture, and blackboard, and of course we didn't have any technology. The most advanced technology we had was the mimeograph machine that we could use to make copies, recalled Professor Wallace.
>
> Our classes were all face-to-face in these large rooms. On average, I had 70 or 80 students, but I think the most I had was 100. That's how all the math classes were at Bordi. There wasn't much interaction at all. I went in and just showed them how to do problems and that was it. There was a midterm and a final and that was it, shared Professor Milacki.
>
> The classes were face-to-face and lecture, but we didn't even refer to them that way. I mean, there was no other modality to compare to, *expounded Professor Mitchell*. I pretty much went to class, took some problems from the textbook, and showed them how to do it, and no I didn't engage them at all. I had no idea if they were learning it or not. In hindsight, it was pretty bad teaching, but I didn't know any better. When

I was hired, my chair gave me the textbook, and said, "Here you go; good
luck." I had no idea what good teaching was all about.

I think I talked from the time class started till it ended, *shared Professor
DeLeon*. It was how I was taught, and I learned math, so I figured that is
how I needed to teach. The funny thing is I have degrees in education,
and I learned different ways to work with students, but I figured that
was [for] elementary school. In college, my math classes were always in
lecture halls. The professors spoke, and we didn't.

However, other participants discussed that even in the 1970s, their institutions were utilizing alternative teaching techniques in developmental and
introductory college-level math classes. As I will discuss in later chapters,
community colleges began using the emporium model of instruction, particularly for developmental math. In general, the emporium model consists
of students learning math in some type of self-paced laboratory setting. In
the late 1970s, Lester Community College utilized this type of instruction for
their developmental courses. Professor Fenimore elaborated:

It was in a classroom but independent study. Students could go at their
own pace. The bright students could work ahead. We had resources like
worksheets for students to practice, and we had a tutor in the classroom.
They got a lot of practice that way. When students finished a unit, they
could take the exam. So, some students finished the class early.

Professor Fenimore explained that there were lecture components in this
modality.

There were common areas where the students got stuck like fractions,
signed numbers, and order of operations, and I would do some short
lectures to the entire class to help them.

By the late 1970s, Sisco Community College adapted a similar emporium
model to Lester's

We just found that the developmental math students were such a heterogeneous group, *recalled Professor Ballard*. You had students who just
had a bad day on the placement test and needed a quick brush-up, and
those students could move through the course quicker, but then you had
students who really lacked basic skills, like arithmetic or basic algebra
skills, and those students needed a lot of drill and practice.

Professor Ballard added:

You have to understand. None of us really understood anything about
developmental math at the time, except that we had to teach it. We all
had math degrees, so we had no experience with developmental math.
There wasn't all the research about it like there is now. We were learning on the job.

Professor Milacki found that for large class sizes with such an academically diverse student population, straight lecture did not work:

> The developmental classes, especially, were just so big and the student skill levels were just all over the place. I had students struggling with adding fractions and finding a common factor and some students just needed two minutes of a review of evaluating an expression and they got it.

This led Professor Milacki to employ more of an emporium model approach.

> I started making up worksheets, bringing them to class and let the more advanced students work ahead. I let them work on problems together in groups. It worked nicely because students wound up helping each other, and it made my job a little easier. Then, the more advanced students could take the tests earlier. It gave me more time to devote to the struggling students.

Professor Milacki elaborated.

> It was a large room, so I had two groups. I moved the more advanced students to one side of the room, and I would have the struggling students on the other. I would have to run back and forth between lecturing to the struggling students and helping the advanced students. Oh, and I had to proctor people taking tests at different times!

Another practice that I will discuss in Chapter 5 is learning communities. Tinto (1998) defined learning communities as "a kind of co-registration or block scheduling that enables students to take courses together. The same students register for two or more courses, forming a sort of study team" (p. 169). Professor Morgan discussed how Habyan Community College implemented learning communities in 1978.

> It [learning communities] had been in the works for a while, even before I got there. Faculty on our campus were attending some national conferences and learning about how it's important for community college students to develop connections with each other. It made sense to me, because they don't dorm and live together like at a university, so it's much harder to make friends.

Professor Morgan elaborated:

> A bunch of faculty from Habyan approached our administration and asked if they could have students take more classes together. For example, the same students taking introduction to statistics would also be taking English 101. Our school was pretty small at the time, so the administration went for it. I remember, the first quarter, I taught an introduction to statistics course and those students were also in a sociology 101 class.

I asked Professor Morgan if the administration set guidelines or requirements for the learning communities.

> Actually, no, I believe it was completely faculty driven. Our school had
> these learning communities coordinators. They were faculty who came
> around and helped us with the learning communities. There really weren't
> any requirements. We were just encouraged to do more activities to help
> students get to know each other. For example, instead of always lecturing,
> we should allow students to work together to solve problems in class.

Did learning communities positively impact students?

> I can't really say, because we didn't keep track of success rates. I can't say
> more students passed my class, maybe a little more. It helped break the
> ice, and it was nice to see students talk to each other, and it broke up the
> monotony of my just lecturing all the time.

It is noteworthy that the origin of learning communities can be traced back
to 1927 when Alexander Meiklejohn designed the two-year Experimental
College at the University of Wisconsin. Students compared Greek literature
to contemporary American literature. Ultimately, the rationale behind the
Experimental College was that undergraduate college studies should help
students to become active citizens with the intellectual skills to participate in
a democratic society (Powell, 1981).

As the decade progressed Professor Ballard discovered how professional
development helped his pedagogical practices

> I think it was 1977 or 1978, my [department] chair asked me if I wanted to
> go to the AMATYC conference. I was like, "What the heck is that?" She
> explained it was where instructors from community colleges met to talk
> about how to teach better. I was like, "OK. What the heck!"

Professor Ballard was impressed:

> It was great. I got to talk with teachers all over the country about stu-
> dents, how we are teaching math, and the common problems we were
> having. In this one session, this one teacher was talking [about] how she
> better organized her work when teaching students. I had never really
> thought about it that way.

Professor Ballard gave an example:

> The teacher [from AMATYC] showed us a clear and organized way to
> explain solving equations with fractions. It's hard to explain, but when
> I taught that in intermediate algebra, I wasn't as organized, and I didn't
> show every step in a clear way. It's like I was showing only three or four
> steps. It had been a long time since I learned solving equations, so I did
> them a lot quicker without really thinking about the steps. I just assumed
> students should get it. But this teacher showed five steps but clear and
> organized. I just learned better ways to explain concepts to students.

Study Guides....What a Concept!

The use of study guides or practice tests, a document with questions from a specific unit that helps students prepare for an exam, is a common practice in the twenty-first century. However, this was not always the case.

> This is going to sound really stupid, but I was talking with my colleagues one day and we were trying to figure how to help students prepare for exams. One of my colleagues said, "What if we could just give our students a sheet with practice problems that focuses on the unit we covered? They could complete it, and we could go over it in class," and the rest of us are like, "Yeah that could work." We tried it, and the students loved it. Please don't ask me why on earth we never thought of that before, because I don't know, recalled Professor Mitchell.

Other faculty members learned about utilizing study guides through AMATYC. Professor Ballard elaborated:

> I was at a concurrent session in AMATYC, and the speaker kept talking about study guides. I raised my hand sheepishly, and asked "What's a study guide?" The speaker was kind enough to explain how he would create a study guide with a bunch of questions from the unit they covered. They are basically practice questions. I was like, "This is a good idea. How did I never think about this?" I went back and told my colleagues about this great idea.

Professor Morgan, who learned about study guides from a colleague, explained:

> It was just another example of how the times were different. When I was in college, my professors lectured, and that was it. No one ever gave me a study guide or a practice test. By the 1970s, we were trying to figure out better ways to not only teach but help our students master the material.

Community College Math Student Population in the 1970s

I asked the participants about the demographics of the community college student population in the 1970s, especially compared to the twenty-first century.

> They were much younger. We didn't have as many of the older and returning students we have now, recalled Professor Ballard.
> It was mostly White students. We didn't have nearly as many Black or Latino students as we do now, stated Professor Milacki.

Were students in the 1970s more motivated to learn mathematics?

> Not really. It was the same stuff as now. Not coming to class as much as
> they should or at all, not studying enough, not doing homework, said
> Professor Morgan.
>   There were still motivational problems when I started, *recalled Professor
> Fenimore*. I could always see the attitude problems in the 18- and 19-year-
> olds on the first day.
>   There will still a lot of unmotivated students, but I feel like the reasons
> were different than now, *explained Professor DeLeon*. Students were just
> lazy and came in with poor study skills and bad work ethics. It wasn't
> like they had the family problems and other major outside problems they
> have nowadays.
>   Students still talked during class, didn't take notes, and put their heads
> down if they weren't interested, added Professor Mitchell.

Did students struggle with basic arithmetic concepts in the 1970s as much
they do in the twenty-first century? Professor Fenimore explained:

> Students still had trouble with fractions and decimals. We had students
> who had only one year or two years of math in high school.

Professor DeLeon shared:

> I was so surprised at how low the students' skill level was. I took the
> job teaching developmental math thinking these students simply needed
> a brush up, but I couldn't believe I was teaching arithmetic. I couldn't
> believe students did not understand how to find the greatest common
> factor or how to multiply fractions or how to add decimals.

Professor Milacki

> I taught an arithmetic class my first quarter, and the shocker for me was
> I had two students who did not know their multiplication tables. The sad
> thing was in all the years I taught developmental math, I kept seeing this.

The Difference between Lower- and
Higher-Level Math Students

As the faculty were learning how to effectively teach their students, and
learning about their students, they were starting to understand the differ-
ences between the developmental math students and the higher-level students. Professor Morgan explained:

> When I started using study guides to review for exams, I decided to try
> something with my college algebra classes. I had them make up their
> own questions to the study guide. I asked them to look over what we
> covered in the unit and come up with questions that they thought they

needed to know for the test. It worked great. They worked together and they created questions using logarithms and hyperbolas, and they came up with great questions. I think creating the questions themselves, rather than just answering them, helped them even more.

Professor Morgan tried this strategy with her basic mathematics class and got different results.

It was a mess. I had students who really didn't understand what a proportion was, so how could they create one? They really didn't understand the difference between an equation and an expression so you wouldn't believe what they came up with for evaluating expressions and solving equations.

Professor Morgan provided another example:

I started to try and engage students more so, I would put wrong answers with the problems worked out on the overhead projector. Obviously, I would block out the student's name. I would ask the class to identify the incorrect steps. My college algebra students could look at complex topics like permutations and combinations or solving polynomial inequalities, these problems with multiple steps, and identify what went wrong. I would show my basic math students the addition of two fractions, and they would sit and stare at me when I asked what went wrong.

Professor Ballard summarized:

Part of the community college learning experience for many of us was that in order to help many of our developmental math students, we had to start from the ground up. It was basically bucket filling with basic information, and they had to understand that basic information before they could apply it.

Learning Disabilities

In 1973, the United States Congress passed *Section 504 of the Rehabilitation Act of 1973*. This act bans the discrimination of people with disabilities in programs that receive federal funding. Public schools, including community colleges, complied. This was new to the community college faculty.

I'm embarrassed to admit, but when I started teaching, I really had never heard of learning disabilities, *stated Professor Ballard*. I hate to admit it, but I thought students who were receiving special services were just lazy and making up excuses.

Some of my students would see a learning disabilities specialist, and I didn't really understand anything about that, *shared Professor Milacki*.

I spoke with one of the specialists and she talked to me about dyslexia, how some students read differently.

If our students provided documentation, they could take their exams in the disabilities office, *reflected Professor Morgan*. They could get extra time or have material read to them. At first I thought it was cheating, so I was practically cross-examining the people in the disabilities office. It was part of my acclimation to the community college population, understanding people with different needs.

Calculator Policy

While adding machines had existed since the 1600s, four-function handheld calculators were invented in the 1960s. In 1972, Hewlett-Packard invented the first pocket scientific calculator, which had more mathematical functions. I asked the participants about the calculator polices in developmental and college-level math courses in the 1970s. All participants mentioned that when they started teaching, there were no calculators used in any math classes.

We didn't use calculators, and quite frankly they were so expensive at the time; students didn't even have them or ask about them, shared Professor Ballard.

For our statistics class, we had one room with five adding machines that students would take turns using, recalled Professor Wallace.

I used to tell my college algebra students how in the olden days students had to use logarithmic and trig tables in the back of the book because we didn't have calculators. They have no idea how good they have it, said Professor Milacki.

Whenever my statistics students complained in the 2000s, I would tell them how my students in the [19]70s used to have to compute standard deviation by hand, commented Professor Morgan.

It was much simpler times, *recalled Professor DeLeon*. Students in the arithmetic course had to do fractions, decimals and order of operations by hand, but it wasn't like the 90s or the 2000s when they were calculator dependent, and we had to teach them without the calculator. They had never used the calculator so there was no arguing about that.

Formal Department Meetings

Department meetings are a standard practice in the twenty-first century. This allows faculty members to congregate and review or modify the infrastructure of the department. Such meetings also allow for professional growth and development. I asked the participants if this was common practice in the 1970s as well. The answers varied:

> We always had [math] area meetings. We had five full-time faculty, and we had a coordinator who called the meetings and determined the agenda, shared Professor Fenimore.
>
> We had meetings, but they were very infrequent. I think we met at the beginning of the schoolyear and at the end of the schoolyear. It was a luncheon where the chairperson just made announcements and went over policies, but we never really discussed anything math related, Professor Ballard.
>
> We had math meetings once a semester, and they were helpful. Something that really helped me was something my chair introduced called a showcase of ideas. Instructors would discuss math topics where students struggle, and we would discuss effective ways to teach them, stated Professor DeLeon.
>
> Our department meetings kind of evolved during the decade. When I first started teaching, we would get together once a year just for announcements and a review of academic policies, but by the end [of the 1970s], we would discuss course content a little more. We would discuss college algebra, for example, and what was working and not working with the course, said Professor Milacki.
>
> We were a tiny department, so we didn't have a lot of meetings, *shared Professor Mitchell*. Our chair disseminated the syllabus, and that was it. If there were announcements or changes, she would pretty much talk to us in the hallway going to class.

Relationships with Administration

As I will discuss, in greater detail, later in this book, a solid relationship between community college math faculty and their administration is an important component for a successful program. Therefore, for a historical perspective, I asked the respondents about the relationship between community college math faculty and their respective administration in the 1970s. Several remembered a very positive and supportive rapport. Professor Ballard elaborated:

> Our division dean and our president used to come to our faculty parties. Both were really supportive and appreciative of the work we did with our students. My president actually knew my name!

Professor Milacki added:

> You have to remember the times were very different. The mission of the community college was to help students who previously didn't have a chance in college. Our president used to talk about that all the time and how we were helping developmental math students, so yes, my dean and president, at the time, really liked what we did in developmental math.

Professors Morgan and Mitchell provided examples.

> When we started the learning communities, our dean came to a meeting. He listened to what we were doing and was very complimentary. He simply told us to tell him if we needed anything, said Professor Morgan.
>
> One time I passed the president in the hall, and he stopped to thank me for the work I did on our placement test. He just went on and on. At the time, I didn't think anything of it, but that would never happen now, shared Professor Mitchell.

However, Professor Milacki felt some frustration with her administration.

> As I said, I had like 70 or 80 students in my classes, and this was really challenging in developmental math because the skill-levels were so varied. I felt like I wasn't teaching them to the best of my abilities because there were so many of them. Every class I was just running around crazy.

Every few years faculty members at Bordi Community College were evaluated by their division dean. This consisted of the dean coming to observe a class. Professor Milacki thought that her dean witnessing her situation would provide relief:

> I was hoping for smaller class sizes. I loved working with these students, but I just couldn't handle the volume of students with so many needs. I figured that when he [Professor Milacki's dean] saw how crazy my class was, he would help.

Professor Milacki was discouraged by her dean's reaction.

> He really didn't see the big deal. He pretty much told me that most of the students will drop out anyway, which was true, so why even start with smaller classes?

Professor Milacki added:

> I was young then, so I didn't understand how community colleges worked. I just taught the students, but I later realized that more students meant more funding for the school. He [the dean] didn't care that these students dropped out. He was happy that Bordi got their tuition money, and the state gave the school money for the students attending.

Summary

While American higher education has existed since 1636, and math has existed in American higher education since the 1700s, math in the community college system developed in the 1970s. Community college math faculty during that time were pioneers in their field. From the 17th through most of the twentieth century, lecture was the primary source of instruction in American higher education. Learning mathematics consisted of lecture followed by drill and practice. This continued to be the primary source of instruction in community college mathematics in the 1970s as well. However, some schools started employing more of an emporium model and even started employing learning communities to focus on student development. While incoming placement exams were nothing new, community college math faculty during the 1970s were instrumental in establishing mathematics placement testing for community colleges. Faculty generally designed in-house exams for incoming students, and in some cases collaborated with faculty from other community colleges to construct such exams.

In the 1970s, faculty contended with some of the same student concerns as the twenty-first century such as attendance issues, poor worth-ethic, and lack of motivation. The participants, however, did provide examples of a positive rapport with their administration. With a foundation set, experiences to build on, and increasing professional development at their disposal, community college faculty were ready to serve students as the twentieth century progressed.

References

Blair, R. M. (1999). *The history of AMATYC: 1974–1999*. Memphis: American Mathematical Association of Two-Year Colleges. https://cdn.ymaws.com/amatyc.org/resource/resmgr/history/amatychistory.pdf

Cohen, A. M. & Brawer, F. B. (2008). *The American community college* (5th ed). Jossey-Bass.

Powell, J. W. (1981). *The Experimental College*. Santa Ana: Seven Locks Press

Thelin, J. R. (2004). *A History of American Higher Education*. Baltimore: Johns Hopkins University Press.

Tinto, V. (1998). Colleges as communities: Taking research on student persistence seriously. *The Review of Higher Education, 21*(2), 167–177.

3

The 1980s: Community College Mathematics Reaches New Heights

As the 1970s morphed into the 1980s, community college enrollment increased. The mission to serve the universal continued, and community college math faculty worked to find more innovative ways to reach their students.

Participants in This Chapter

In this chapter, Professors Ballard (Sisco Community College), Milacki (Bordi Community College), Mitchell (Telford Community College), Morgan (Habyan Community College), DeLeon (Griffin Community College), Fenimore (Lester Community College), and Wallace (Lester Community College) will share their continued experiences teaching community college math. There will also be three new participants who started teaching in the 1980s:

Professor Thurmond

Professor Thurmond taught developmental math, full-time, exclusively from 1980 to 2015 at Griffin Community College where developmental math and college-level math were in separate departments. She served as chair of developmental education from 2006 to 2012. Professor Thurmond has a bachelor's and master's degree in education.

Professor Bell

Professor Bell taught both developmental and college-level math at Habyan Community College from 1983 to 2018. Professor Bell served as chair of the math department from 1988 to 2012. Professor Bell has a bachelor's and master's degree in math. He has a Ph.D. in math education.

DOI: 10.1201/9781003287254-3

Professor Johnson

Professor Johnson instructed both developmental and college-level math, full-time, at Kilgus Community College from 1989 to 2012. She has a bachelor's degree in math, a master's degree in statistics, and a Ph.D. in education.

Continued Professional Development

AMATYC (American Mathematical Association of Two-Year Colleges)

AMATYC continued to grow and flourish. By 1983, its membership grew to 951 individual members and 78 institutional members. In addition to the Mathematical Association of America and the National Council of Teachers of Mathematics, AMATYC had become one of the major national organizations of mathematics. In 1987, AMATYC membership reached 1,500 members. The foci of AMATYC by the 1980s consisted of the employment of academic computers in the classroom, the developmental math curriculum, professional trainers for two-year math faculty, and the encouragement of women in mathematics. AMATYC continued to gain more notoriety, as the first international conference was held in Calgary, Canada (Blair, 1999).

NADE (National Association for Developmental Education)

Professional development focusing on the understanding and improvement of developmental education commenced in the late 1970s. This work began in 1976 when a small group of community college and university faculty began gathering in Chicago to discuss their work in remedial education. At this point, remedial education was being referred to as developmental education. Rather than reteaching content from elementary and secondary education, there was more of an emphasis on student development (Boylan, 2016). In 1976, a formal organization focusing on developmental education was funded by the W.K. Kellogg Foundation at Appalachian State University (Spann, 1996). The organization was originally titled, the National Association for Remedial/ Developmental Studies in Postsecondary Education (NAR/DSPE). In 1984, the organization's name was changed to the National Association for Developmental Education (NADE) (Boylan, 2016). In 1978, the *Journal of Developmental and Remedial Education* began to produce scholarly articles focusing on developmental education.

The discipline of developmental education continued to grow. In 1980, NAR/DSPE (later NADE) established the Kellogg Institute of Training and Certification of Developmental Educators at Appalachian State University. This was America's original professional development program for developmental education (Spann, 1996). In 1984, *the Journal of Developmental Education,* which contained empirical and conceptual research geared toward improving developmental education, commenced (Boylan, 2016). *Research and Teaching in Developmental Education,* another journal focusing on developmental education, followed in 1985. In 1984, developmental education became recognized as a national discipline as the National Center for Educational Statistics published its first report on developmental education. Furthermore, in 1986, Grambling State University, in Louisiana, founded the nation's first doctoral program in developmental education. (Boylan & Bonham, 2007).

Course Structure (Circa Mid-1980s)

Below are content descriptions for the developmental and introductory math courses at the community colleges, in this study, during the mid-1980s. While the mathematical content remained stagnant, there was some notable shift in the course structure for the arithmetic and algebra courses. I did not list the introduction to statistics and college algebra courses for the 1980s, as there were no major changes from the 1970s. The only change was most schools shifted the topics of permutations and combinations from college algebra to statistics, and the binomial theorem was dropped from most college algebra classes. Habyan Community College also added a trigonometry component to their college algebra class. Kilgus Community College is omitted, as their records were not available.

Sisco Community College (Quarter System)

Basic Mathematics (Developmental Math—No Credit)

This class focused on all arithmetic operations (whole numbers, fractions, decimals, percentages), units of measurements, and operations with integers.

Algebra 1 (Credit-Bearing Course)

This course emphasized operations with rational numbers, simplifying algebraic expressions, literal expressions, linear algebraic equations, laws of exponents, and factoring polynomials.

Algebra 2 (Credit-Bearing Course)

This class consisted of laws of exponents, factoring polynomials, operations with rational expressions, system of linear equations, graphs of linear equations.

Algebra 3 (Credit-Bearing Course)

Algebra 3 focused on quadratic equations, complex numbers, inequalities, roots and radicals, numbers, parabolas, and absolute value equations.

Telford Community College (Quarter System)

Computational Skills (Developmental Math)

Computational skills emphasized operations with whole numbers, fractions, decimals, prime factorization, ratio and proportions, percentages, and operations with rational numbers.

Basic Concepts of Algebra (Developmental Math)

This course reviewed operations with fractions and decimals, and emphasized operations with rational numbers, evaluating expressions, linear equations, solving inequalities, operations with polynomials.

Elementary Algebra (Credit-Bearing Course in the Math Department)

This course highlighted linear equations, operations with polynomials, laws of exponents, factoring, rational expressions, roots and radicals, and graphs of linear equations and inequalities.

Intermediate Algebra (Credit-Bearing Course in the Math Department)

This course focused on system of linear equations, quadratic equations, parabolas, complex numbers, roots and radicals.

Bordi Community College (Quarter System)

Basic Concepts of Arithmetic (Developmental Math)

This course focused on operations with whole numbers, fractions, decimals, ratios and proportions, units of measurements, and operations with rational numbers.

Fundamentals of Algebra 1 (Credit-Bearing Course)

This class emphasized operations with rational numbers, evaluating expressions, solving linear equations, problem-solving applications, solving linear equations, graphs of linear equations, solving laws of exponents, factoring.

Fundamentals of Algebra 2 (Credit-Bearing Course)

Highlighted in this class were evaluating algebraic expressions, solving linear and quadratic equations, operations with rational expressions, and parabolas

Fundamentals of Algebra 3 (Credit-Bearing Course)

This class emphasized inequalities, simultaneous equations [system of equations], roots and radicals, absolute value equations, complex numbers, and problem-solving applications.

Habyan Community College (Quarter System)

Basic Mathematics (Developmental Course)

This course emphasized operations with whole numbers, fractions, decimals, percentages, proportions, units of measurement, and signed numbers.

Introduction to Algebra (Developmental Math)

Highlighted in this class was a review of arithmetic skills, operations with rational numbers, evaluating expressions, linear equations, laws of exponents, and factoring.

Elementary Algebra (Credit-Bearing Course)

This class focused on operations with rational numbers, solving equations, literal expressions, laws of exponents, factoring, rational expressions, and operations with polynomials.

Intermediate Algebra (Credit-Bearing Course)

Highlighted in the course were quadratic equations, simultaneous equations, complex numbers, roots and radicals, inequalities, relations and functions.

Griffin Community College (Semester System)

Basic Arithmetic (Developmental Course)

This course highlighted operations with whole numbers, fractions, decimals, proportions, units of measurements, percentages, and operations with integers.

Algebra 1 (Credit-Bearing Course in the Math Department)

This course focused on operations with integers and rational numbers, evaluating expressions, linear equations, system of linear equations, laws of exponents, factoring, and operations with rational expressions.

Algebra 2 (Credit-Bearing Course in the Math Department)

Emphasized in the class were operations with polynomials, quadratic equations, roots and radical equations, complex numbers, inequalities, absolute value equations, relations and functions, and applications.

Lester Community College (Quarter System)

Fundamentals of Arithmetic (Developmental Mathematics)

This class focused on whole numbers, fractions, decimals, common fractions, metric measurement, ratios, proportions, and percentages.

Introduction to Mathematics (Developmental Mathematics)

Highlighted in this class were beginning geometry and pre-algebra topics focusing on number system operations, and interpretation of algebraic symbols.

Elementary Algebra (Credit-Bearing Course in the Math Department)

This course emphasized operations with signed numbers, laws of exponents, literal expressions, polynomials, first-degree equations, factoring, and algebraic fractions.

Intermediate Algebra (Credit-Bearing Course in the Math Department)

This class emphasized sets, real numbers, polynomials, algebraic fractions, first degree equations and inequalities in one variable, radical expressions, complex numbers, quadratic equations in one variable, graphs in the plane, system of linear equations, relations, functions, proportions, logarithms.

Changes in Course Structure

When comparing the community college course content from the 1970s to the 1980s, the most notable change was that courses that consisted of both arithmetic and introductory algebra content were split into additional courses, thereby lengthening the developmental math sequence. I asked the respondents the rationale for this.

> Students were so deficient in arithmetic skills. They couldn't get fractions or decimals. Some struggled with whole numbers, multiplication and division, and you can't do algebra if you don't understand arithmetic, so we developed another course, explained Professor Ballard.
>
> There is a reason arithmetic is taught over several years in elementary school. Learning fractions, decimals, and percentages takes time. Some [students] needed a quick brush-up, but many needed deep teaching for those topics. They needed more time, added Professor Milacki.
>
> We had students who did not understand how to multiply fractions trying to evaluate algebraic expressions with fractions or trying to solve equations with fractions. Obviously, they couldn't do it. Students needed more time and help with arithmetic, but I think adding the second course [Introduction to Algebra] was about establishing a break point. Students had to demonstrate competency in arithmetic before they could sign up for algebra, said Professor Morgan.

Did lengthening the sequence lead to better success rates in the arithmetic and introductory algebra courses?

> Maybe a little, but not a lot. Nothing noticeable, *recalled Professor Mitchell.* We were just trying to be fair to the students. So many of them needed so much help in fractions and decimals that we thought we were giving them more of a chance to succeed.

Was the administration supportive in adding an extra course?

> As long as enrollment was high, which it was, and students stayed long enough to be counted in the census, our administration was happy, and they didn't care what we did to our courses, said Professor Morgan.

Another small, but consistent, change in community college course curricula was the shifting of permutations and combinations from college algebra to statistics. Professor Ballard explained:

> We just realized they [permutations and combinations] were just a better fit for an introductory to statistics course. Factorials are used in higher level math, so I guess that's why they were taught in college algebra, but our students were struggling in our statistics sequence without having permutations and combinations taught earlier.

Professor Ballard added:

> It was another example of the benefits of us, as faculty, just talking more about our classes. We didn't do that enough when I first started, but we did that more as time went on.

A Shift to Standardized Placement Testing

In the 1970s, community colleges generally employed locally generated placement exams. More specifically, such exams were tailor made to assess the students' aptitude for the individual community college's math classes. Some community colleges did not even offer a placement exam. However, in the 1980s this began to change. Administrators and educators began to push for mandated placement testing. Florida, Georgia, Texas, California, and New Jersey were among the inaugural states to implement mandatory placement testing (Cohen & Brawer, 2008). More specifically, due to the difficulty students faced in mathematics, researchers began to recommend placement testing for students (Wood, 1980). By the mid-1980s, over 90% of American community colleges were employing some sort of incoming placement exam (Cohen & Brawer, 2008).

The types of placement exams also changed in the 1980s. In contrast to locally designed paper and pencil placement exams, community colleges began to utilize nationally standardized and computerized placement exams. The College Board developed Compass in 1983 and ACCUPLACER in 1985. Students took both exams on the computer, and their results were electronically assessed. While these exams still had cut scores, to get into specific classes, they were computer adaptive in that a student's response for one question dictated the next type of question he or she would receive. While Compass and ACCUPLACER were much easier to administer and assess, the national standardization and adaptiveness meant much less control for faculty. By the conclusion of the 1980s, all the community colleges in this study were employing either Compass or ACCUPLACER. Faculty responses were mixed. Professor Morgan elaborated.

> I get it. It was a lot of work to come up with and keep updating those placement exams, and as our enrollment grew, I'm sure it became more demanding to grade them by hand, but I didn't like the loss of control we had with the new placement exams. For the first time, I didn't know what questions my students had to answer to get into my classes. To get a better idea, I took the test myself, but because they are adaptive, I still had no idea what specific questions my students were getting.

Professor Wallace shared:

> I was quite perturbed when we went to a computerized, standardized test. It just made no sense to me to use a test that is kind of off-the-shelf, designed for every college in the world.

Professor Wallace added:

> Yes, the curricula are pretty similar from college to college, but not exactly so. Some topics [on the test] may be taught before other topics in other schools. So, it just always made more sense to have a placement test catered to your individual curriculum.

Professor Ballard recalled how this negatively impacted placement testing.

> I remember we had more improperly placed students; they were less prepared. I had more students in algebra who could not do arithmetic. I had more students in intermediate and college algebra who had trouble with basic algebra.

Professor Ballard offered an explanation:

> Some of the questions were multiple choice, so they can guess. That means they can get in higher math without getting the correct answers.

Lester Community College began employing Compass, and Professor Wallace also noted a change in student placement; however, it was the opposite of what Professor Ballard noted.

> We had more students testing into developmental math (fundamentals of arithmetic, introduction to algebra) and less students testing into elementary algebra. The strange thing was our success rates in elementary algebra went down.

Like Professor Ballard, Professor Wallace had some thoughts on this occurrence:

> I think it [the placement test] was testing all the wrong things. It's a standardized test. If the hierarchy of topics on the test doesn't agree with the hierarchy of topics in the course then you get all kinds of mismatches.

Improving Student Engagement

In the 1980s, students continued to struggle in math. While community colleges did not keep track of student success rates, the faculty understood their success rates were poor.

> It was tough to see half the class, maybe more, fail or drop out. We all wanted to do more to help our students, Professor Mitchell.

While lecture-based instruction remained the primary teaching modality throughout the 1980s, faculty employed more student engagement to help their students. Professor DeLeon elaborated.

> I went to an AMATYC conference, and they were talking about if we get more students engaged in our classes, they have a better chance to learn. I learned about the think-pair share practice. I noticed students became more involved in my classes.

Professor DeLeon provided an example of this technique.

> I would give students a problem but then have them compare and discuss answers in pairs. It was a big help in a lot of ways, but in many cases, students helped each other out.

In Chapter 2, Professor Morgan discussed the use of learning communities at Habyan Community College in the 1970s. However, by the mid-1980s, they began to dissipate.

> As our enrollment grew, it became harder to schedule blocks of time where students could register for classes together. We started getting more non-traditional and older students who worked more, and that made scheduling harder.

However, Professor Morgan continued to employ the practices from her use of learning communities.

> I kept using group work. After introducing a topic, I would assign groups and have students work on the problems together. In many cases, I would have groups of students explain their answers to the class. It just kept everyone involved.

In her book *Student Engagement Techniques: A Handbook for College Faculty*, Elizabeth Barkley (2010) discusses various techniques that college faculty can employ to engage students. However, in the early 1980s, Professor Morgan began using a variation of the technique, small group tutorials. She explained:

> When it came to troublesome topics such as fractions, signed numbers, or solving equations, I would identify the students who were strong in these areas and those who were struggling. There would be specific days where I would assign groups, and the group leader was obviously the stronger student. The leader would then work with students on assigned problems.

Professor Morgan discussed the benefits of this:

> I just found many students were more comfortable asking questions in small groups and getting help from their peers. It was an eye opener for me. I never knew when I was doing straight lecture how many students probably had questions but were afraid to ask.

Professor Bell discussed the how even the slight employment of various techniques can increase student engagement.

> I learned about wait time. It's where you ask a question, but you give your students time to answer, and you pay more attention to their expressions to get an idea if they are understanding.

Professor Bell continued:

> I also learned about how brain-breaks and movement helped students. So, during class, after doing a few problems, I would have students get up and move to a different part of the room. It helped break things up and kept them from getting bored.

Professor Bell explained:

> You have to understand. I never took a single education course, so when I started teaching, I would just show up and teach, but I learned through going to AMATYC and working with my colleagues, that I could be doing more. I know all this sounds simple now, but it was cutting edge, at least for higher education, back then.

Preparation Is the Name of the Game

In the 1980s, faculty were still learning about the art of becoming effective community college teachers. Professional development and collaboration with peers were imperative to this growth; however, they learned some lessons by experience. The faculty learned that sufficient and even extensive preparation was salient to quality instruction.

> I thought teaching was easy. I just showed up and did some sample problems. At times, I would find myself struggling with explanations and having to think through the problem too much on the fly. That was an indication for me that I needed to prepare more in advance the problems I was going to do, shared Professor Bell.
>
> I have a master's in math but when I started teaching statistics and college algebra, I'm embarrassed to admit that there were problems that I would draw blanks on when doing them in front of the class, admitted Professor Mitchell.
>
> I'll never forget it. This one time I was teaching word problem applications in algebra and I was stumbling with the setup, and one of my students actually called out, "Could you practice some of these problems before you teach them." Yes, that was my wake-up call, shared Professor Thurmond.
>
> Something I learned from working with especially developmental students is that meticulousness is very important. The slightest bit of ambiguity in a problem can throw off a student. It could be the tiniest little thing, shared Professor Milacki.

I asked Professor Milacki to share an example:

> In long division of polynomials, many students get confused in the third step when you have to subtract the terms and contend with the signed numbers. The bottom line is thorough preparation can help you help students with these specific details.

Professor Ballard discussed how preparation can help a teacher to expect the expected:

> As a new teacher, I struggled sometimes with students' questions. Many times, they blindsided me. I started keeping track of common questions and common struggles students had with certain topics. I would review this before teaching a topic. This helped me anticipate and answer questions better.

Professor Deleon shared how preparation can lead to better efficiency

> I struggled with repeating myself too much and taking too long to explain a problem. Better preparation helped me to be cut down on some of the repetition.

A Slightly Changing but Still-the-Same Student Population

The Participants mentioned that the student population for community college math remained largely the same in the 1980s.

> We were still getting them right out of high school, lots of 18- and 19-year-olds, recalled Professor Fenimore.
>
> They still struggled with all the same concepts, *reflected Professor Mitchell*. As I got more experienced, I knew the trouble areas before I taught them. I knew they would struggle with fractions in any class. In algebra it was always the signed numbers, factoring, rational expressions, system of equations, especially the three by three. They would struggle with completing the square as well. It became timeless.

However, some participants noted a change in the student population as the decade progressed.

> I noticed we were getting more non-traditional students, especially people in their 30s and 40s coming back to school. It was gradual, but it was noticeable, reported Professor Ballard.
>
> We started getting more students who already had children. Some were married with children, and some were single parents, *recalled Professor Milacki*. As a mother myself, that amazed me, that they could go to school while raising children. You know something? At the time, they [students who were parents] were my best students, because they took school seriously.

Professor DeLeon noted this as well and offered a theory:

> I think it had to do with job competition. By the [19]80s, more jobs wanted more education and either associate or bachelor's degrees. In the late [19]80s, companies and employers wanted computer skills, so that sent people back to school. And of course, math is required for every degree, so we got them all.

Professor Thurmond added:

> I think the open access community college was an eye opener. People looking for a higher quality of life had a chance to get one through education. For example, I started getting more women who were recently divorced. They got married right out of high school, and they really didn't have any skill, but now they needed to get a job, and they wanted a good job. In the past, going back to school to get a better job wasn't an option for a lot of people.

The Birth of Tutorial Services

To increase student success, the faculty searched for ways to assist students outside the classroom. Consequently, in the 1980s, they initiated math tutorial centers and math help rooms. Professor Ballard expounded:

> My colleagues and I were traveling to conferences like AMATYC and NADE to find ways to help our students. I can't remember if we got the idea from AMATYC or NADE or both, but we heard about math tutorial centers, separate rooms where students could get extra help with their math during the day, so we decided to set that up.

Professor Ballard continued:

> As a department, we wrote out the proposal, and our administration approved. We basically just converted a classroom into the math help room. It was around 1982–1983 when it opened. Students could come in, complete worksheets, and just get help with their math homework. We hired some student tutors, and we hired a coordinator to run the place, but mostly it was the math faculty. We volunteered our time to help students and make sure the place was running well.

In 1984 Griffin Community College opened a math learning center like SCC's. Professor Thurmond explained how they employed technology to help their students.

> Griffin [Community College] was always ahead of the game in technology. VHS tapes had been out for a few years, and we got the idea to record short lessons for students on topics like solving linear equations, operations with polynomials, foiling and factoring, you know the tough topics. Students could come to the math help center and watch these videos in addition to getting individual assistance.

Professor Thurmond discussed the logistics of this.

> We had a state-of-the-art IT [Institutional Technology] team. All I had to do was come in and teach the lesson to the camera, and they converted it to video. Don't ask me what they did.

Lester Community College implemented a math help room, and Professor Fenimore discussed how this made testing easier, especially since they used the emporium model for developmental math classes.

> We didn't have to use class time for exams. We would give students a pass, and they would take their exams in the math learning center. Then, their instructors would grade it.

Bordi Community College opened a math tutorial service center in 1983, and this provided relief to Professor Milacki, who had been utilizing a self-paced emporium model.

> Do you have any idea how hard it was to run a class where some students are testing and some are trying to learn at the same time? It made my job so much easier to just send those students to the math center to take their exams. It made me a better teacher, because I wasn't trying to proctor exams and teach.

If the inception of math learning centers was successful, how did the faculty quantify such success?

> Our math help room was mobbed, *said Professor Bell*. We always had students lined up for tutoring, and they kept coming back. Because our numbers [students receiving tutoring] were so high, the administration supported us moving the math help room to a bigger place and hiring more personnel to help our students.

Emphasis on Study Skills

In the 1980s, the importance of student study skills surfaced. Professor DeLeon explained:

> Don't misunderstand me; I understood that students needed to study and do homework to be successful, but when I entered college, I just knew how to be a good student, but we had a lot of students who came to community college who did not know how to be good students.

Professor DeLeon elaborated on how she emphasized study skills in her classes.

> I started talking about organized note taking, and in my developmental math class, I started walking around the room and getting on their cases if they either weren't taking notes or taking bad notes. I also started talking about the number of hours they should be spending outside of class doing homework and reviewing their notes.

Professor Mitchell also focused on notetaking.

> I started collecting their notes and making comments. What I would do is black out their names and put different students' notes on the overhead projector, and as a class, we would discuss what was good and bad.

Professor Mitchell added:

> I'm not sure you could do that now, because students would probably complain and there are so many privacy issues, but it worked well. Note taking is important, and they got to see examples of the good, the bad, and the ugly.

It was professional development that led Professor Mitchell to this practice.

> I was really interested in helping the developmental math students, so I attended a NADE conference. Man, it was like meeting kindred spirits, all these people who worked with the same kinds of students that I did. But yes, I got that idea from a colleague at NADE.

In the 1980s, incoming freshmen were required to enroll in a basic study skills course. Professor Ballard elaborated:

> I remember when Sisco [Community College] started that class. I actually taught it a few times. They [SCC's administration] wanted faculty who taught freshmen classes to get involved. We talked about proper note taking, reading a textbook, and ways to retain information. You know, the stuff they didn't do.

Counseling for Students

Over time, faculty became aware that students had needs that went beyond mathematics. Therefore, in the 1980s, community colleges began offering counseling services within their departments. Professor Thurmond elaborated:

> I think it was in 1984, our department got our first academic counselor. She would help the students map out their academic plan, but she would also help, especially, the developmental math students with their work habits and study skills.

Telford Community College hired a departmental academic counselor as well in 1985. Professor Mitchell elaborated:

> It got to the point where we would refer our students to him [the counselor] if they were having trouble. If we saw students weren't doing their homework or just seemed disengaged in the class, we would refer our students to the counselor.

Professor Mitchell added:

> I was never sure what he [the counselor] said to the students, but I think it just showed our students that someone cared, and that helped get them back on track.

Professor Milacki shared how Bordi Community College began employing an early alert system in the late 1980s for students.

> By the late 1980s, we had two counselors for our department. Around Week 3, we were asked to create a list of students who were struggling in our classes and make some comments, like if they were failing exams or not doing homework or not attending. The counselors would call the students and just talk to them, see if everything was all right and talk to them about how they could improve.

I asked Professor Milacki how faculty sent these lists to the counselors.

> It was all paper and pencil. This was before computer systems were networked. We would make a list, write the comments and walk downstairs and place the list in the counselor's mailbox.

Professor Mitchell shared the motivation behind bringing academic counselors into their respective departments.

> All over campus faculty were frustrated that students were so unprepared and had so many needs. I was on faculty senate, at the time, so I remember the conversations. We approached the administration about getting a little more help for our students.

Method of Instruction

Throughout the 1980s, the participants emphasized that their pedagogical practices evolved from the 1970s or even early 1980s. Professor DeLeon elaborated:

> I just became more of a well-rounded teacher. By the end of the 1980s, I wouldn't say I was really lecturing. It was more of a guided practice. When explaining a topic, I would involve and call on students. I used group-work and just found ways to get them more involved in class. I guess you could say I became more of a personable teacher.

Professor Ballard concurred:

> I think I speak for most of my colleagues at Sisco [Community College], but we learned how to be community college professors. When I first started teaching, I taught like I was a Harvard professor. I came to understand that these students didn't come to college with the best work ethic. That meant teaching students how to take notes, how to study math, and in some cases how to behave in a college classroom.

In the 1970s, institutions used various forms of the emporium model. This continued and evolved in the 1980s.

> I actually started the emporium model at Bordi [Community College], *recalled Professor Milacki*. I talked with my chair and my colleagues about how I allowed students to self-pace. By the late 1980s, our developmental

math classes were in labs with TV screens and VCRs. Students would watch videos, with headphones, on various topics like finding the least common multiple of two numbers or finding the percent of a number. Then, they would answer questions on a worksheet. After completing various topics, they would take a test in the math help center. By that time, I had smaller class sizes, maybe 30 or 40 students, and I even had one or two student-tutors to help me.

Was there lecture in this model?

There was some. Sometimes, I would stop everyone and go over a topic for about 15 or 20 minutes. It was usually for troublesome topics like adding and subtracting signed numbers.

I asked Professor Milacki how Bordi reduced the class sizes.

I just kept bugging my department chair who kept bugging my dean that we needed smaller class sizes. With enrollment going up, they hired more faculty. As long as our enrollment was high, the administration was happy.

Lester Community College continued to offer developmental math classes in the emporium model as well.

It was very flexible, *said Professor Fenimore*. Students had to complete worksheets and take exams once they were done with a unit, but if they weren't ready for the test, they could have more time. They just had to finish their work by the end of the quarter. Our classes were simply pass or fail. The students just needed a minimum of 80% on each exam.

Kilgus Community College offered developmental math students, as well as those in elementary algebra, intermediate algebra, and statistics, a choice regarding the instructional modality. Students could enroll in a class with lecture-based instruction or a class that employed the emporium model. Professor Johnson explained:

The traditional classes had about 40 students, but the other classes [emporium model] had some 80 or 120 students in it. The class was part lecture, and then you broke into three groups of 40 for two hours a week.

Professor Johnson elaborated on the computer lab:

The computer lab had 40 Apple II's, and students worked on the computer and did drill problems based on the lecture. Students had a floppy disk and did math problems on the computer. This was before computers were networked, so they would print off their scores and give me the results.

In the computer lab, Professor Johnson's role was that of a facilitator.

As they were doing the problems, I would walk around to make sure they were doing the problems, and I would help them if they got stuck.

Technology for Community College Math Courses in the 1980s

While the participants have referenced various forms of technology in this chapter, I wanted to get a specific idea of how technology evolved throughout the 1980s.

Calculators

By the end of the 1980s, the scientific calculator had become widely used in community college math courses. This was quite a change from the beginning of the decade.

> In the [19]70s and even the early [19]80s, all we had was a bunch of adding machines and by 1989, our college algebra and statistics classes were using the scientific calculators, *recalled Professor Mitchell*. It happened pretty fast, but I guess it had to do with the cost. When the cost for basic [four function] calculators came down, we started using them, and the same for scientific calculators.

Professor Ballard discussed the challenges for faculty.

> Using the basic calculators was a piece of cake, but a lot of us had trouble with the scientific calculators. I know a guy who retired just because he couldn't understand how to use them.

Professor Ballard shared his own experience.

> I was assigned to teach statistics in the fall of 1988, and that is when we started using the scientific calculators. I was scared; I had no idea what I was doing, but our faculty really worked together. We had this young guy who knew those things [scientific calculators] cold. He did these workshops over the summer and worked with me individually as well. Our department worked so well together, and everyone helped everyone.

Was the use of the scientific calculator difficult for the student to master?

> Yes, it was, *recalled Professor Bell*. At that time, most of the high schools weren't using scientific calculators, so we had to use them. I taught statistics with the scientific calculator in 1989, and it felt like I was teaching both a math and a computer course at the same time.

Professor Bell explained the rationale for upgrading to the scientific calculator:

> I think we were just trying to stay on the cutting edge. One of our faculty members attended AMATYC and learned about how using the scientific calculator can make learning statistics a little easier, so we figured we would give it a try. I'm glad we did.

Regarding the use of a basic calculator for arithmetic, introductory algebra, and intermediate algebra courses, community colleges were mixed.

> In the 1980s, there were no calculators for any arithmetic or algebra classes, recalled Professor Milacki.
>
> We didn't use the calculator for our basic concepts of arithmetic or any of the basic algebra classes, and we didn't even think about it. High schools at that time still weren't really using the calculators, so it was never an issue, said Professor Bell.

However, by the 1980s, some were considering a change.

> By about 1988 or so, we decided to try the basic calculators for certain topics in our classes. We would use them for word problems only in the arithmetic classes and heavier problems like equations with fractions in algebra. We used the calculator a lot in intermediate algebra, shared Professor Bell.

Professor Ballard elaborated:

> We had it so some exams or certain parts of exams were calculator only. So, for the basic math class, they would turn in part 1 of the exam, which was all computation, but for part 2, which was word problems, they could use the calculator.

Professor Ballard added:

> We got the idea at AMATYC. Some folks in their presentations were talking about how real-life math is messy and requires the use of calculators. It made a lot of sense, so we gave it a shot.

Videos

With the increased use of VHS tapes, short mathematics videos became popular in the 1980s. Earlier in this chapter, Professors Thurmond and Milacki discussed the use of VHS tapes. Professor Thurmond took part in designing videos on VHS tapes at Griffin Community College. Bordi Community College acquired the VHS tapes from an outside source.

> I don't remember the name of the distributer, but we acquired the tapes from somewhere. Between going to AMATYC and talking with faculty at other schools, we heard about good companies that produced that kind of material, reflected Professor Milacki.

Habyan Community College also employed math videos on VHS tapes at their math tutorial center. Professor Morgan explained:

> This [math videos] was new and exciting in those days. It was nice to have short and concise videos, you know ten or fifteen minute videos that get right to the point with a couple of examples. The students really liked them.

The Emergence of Computer Software

The use of desktop computers surged throughout the 1980s in education. However, it was not until the late 1980s when community colleges began implementing content-specific computer software to assist students. Earlier Professor Johnson discussed how students would complete drill and practice through the Apple II computers. However, it was rare that computers were utilized in math classes.

In the 1980s, computers were becoming common in education; however, there was a form of segregation. Rather than computers being integrated into the classroom curriculum, they were supplemental entities. Professor Thurmond elaborated:

> We had a couple of computers in our math tutorial center. I believe they were Apple, and students could practice their math on them. I recall we had a program called Success with Algebra that was popular with some students. But the computers were restricted to the math tutorial center, and most students or faculty didn't really want anything to do with them.

Professor Bell concurred:

> We had some computers in our math help room, and there was some interest among the students. When I was in there, I would see students working on the computers, but I had no idea what they were doing. Our young tutors were also interested and good with computers. But it was a learning curve for most people. I remember a young tutor telling me that computers were a big part of our future in the classroom, and I thought "yeah right."

Professor Morgan discussed how the computer as a special or segregated entity was common in education in the 1980s.

> I had two daughters who were in elementary and middle school in the 1980s. They would go to a computer class once a week. It was like gym class or music class. None of their other teachers incorporated them into the general curriculum.

Continued Positive Relationships with Administration in the 1980s

I asked the participants about their relationships with their respective administrators in the 1980s. In general, the faculty reported a positive rapport.

> It felt like my dean and even the college president were more like our peers, not our supervisors. We didn't always agree on everything, but they really understood our students, the kind of challenges we faced with students, and the kinds of students we had, reflected Professor Ballard.

Let me give you an example of the kind of administration we had, *exclaimed Professor Mitchell*. I got my Ph.D. in 1986 and when I did, my dean, provost and even my president came to my office to congratulate me. I actually thought that was normal back then.

I was frustrated with large class sizes when I first started, but after a while, I felt like they [the administration] listened to our needs a lot more. We had lower class sizes, better facilities like the math help room, and counselors for our department, *said Professor Milacki*. But developmental math, with our high enrollment, was a huge cash cow. I guess they didn't want to bite the hand that fed them.

I didn't realize this at the time, but we were respected as true experts in our field. Sometimes my dean would have a question about our math program, and she truly respected our responses and what we did, recalled Professor Bell.

Summary

Throughout the 1980s, both developmental math and introductory college math courses continued to develop. National organizations such as AMATYC and NADE grew, which allowed faculty to cultivate their pedagogical repertoire. Faculty also emphasized study skills to assist students, and community colleges established math learning centers and tutorials to help students outside of class.

Community college placement exams underwent a major transformation in the 1980s. Prior to the 1980s, incoming placement exams were developed internally by faculty at specific institutions, and those exams were tailored to the courses at those colleges. In the 1980s, many community colleges adopted ACCUPLACER and Compass, two computerized placement exams that were nationally standardized and adaptive.

The mathematical course content, for the community colleges, in this study, did not change drastically from the 1970s to the 1980s. However, recognizing that students were deficient in basic arithmetic skills, most schools extended their developmental math sequence to allot more time for students to study arithmetic concepts.

The use of technology in education grew immensely in the 1980s. In the early 1980s, most community colleges were not even using basic calculators; however, by the late 1980s, some were using scientific calculators. Community colleges were also employing VHS tapes as part of math tutorials and even in the math emporium model. Content-specific computer software was even starting to emerge at the 1980s concluded.

The 1980s were a positive and productive time for community college math courses. With a growing wealth of professional development, the rapid growth of technology, support from administration, as well as support among faculty, the future looked bright regarding community college math instruction.

References

Barkley, E. F. (2009). *Student Engagement Techniques: A Handbook for College Faculty*. Jossey-Bass.

Blair, R. M. (1999). *The History of AMATYC: 1974–1999*. Memphis: American Mathematical Association of Two-Year Colleges. https://cdn.ymaws.com/amatyc.org/resource/resmgr/history/amatychistory.pdf.

Boylan, H. (2016). History of the national association for developmental education: 40 years of service to the field. https://thenoss.org/resources/Documents/History/History%20of%20the%20National%20Association%20for%20Developmental%20Education.pdf.

Boylan, H. R. & Bonham, B. S. (2007). 30 years of developmental education: A retrospective. *Journal of Developmental Education, 30*(3), 2–4.

Cohen, A. M. & Brawer, F. B. (2008). *The American Community College (5th ed)*. Jossey-Bass.

Spann, M. G. (1996). National center for developmental education: The formative years. *Journal of Developmental Education, 20*(2), 2–6.

Wood, J. P. (1980). Mathematics placement testing. In F. B. Brawer (Ed.), *New Directions for Community Colleges: Teaching the Sciences*, no. 31, Jossey-Bass.

4

The 1990s: Mathematics Enters a New Age

The 1990s looked promising for the continued growth of community college math. While many students were still struggling, faculty were finding new and innovative ways to help their students. In this chapter, the participants from Chapters 2 and 3 shared their experiences from the 1990s. This included Professors Ballard (Sisco Community College), Mitchell (Telford Community College), Milacki (Bordi Community College), Morgan (Habyan Community College), DeLeon (Griffin Community College), Fenimore (Lester Community (College), Wallace (Lester Community College), Thurmond (Griffin Community College), Bell (Habyan Community College), and Johnson (Kilgus Community College). Additionally, three new respondents conveyed their experiences as well.

Professor McDonald

Professor McDonald taught developmental math full-time at Griffin Community College. He started in 1994 and was still teaching in 2021. He has a bachelor's degree in elementary education and a master's degree in education. When Professor McDonald started, the average class size was 25 to 30 students.

Professor Timlin

Professor Timlin began teaching both developmental math and college-level math at Griffin Community College in 1992. He holds a bachelor's and a master's degree in mathematics.

Professor Mesa

Professor Mesa taught developmental math full-time at Telford Community College. She started in 1993 and retired in 2019. She possesses a bachelor's

DOI: 10.1201/9781003287254-4

degree in elementary education and a master's in counseling. The average class size was approximately 30 students when Professor Mesa began teaching at Telford Community College.

Professional Development Grows and Gains More National Attention

AMATYC (American Mathematical Association of Two-Year Colleges)

The development of AMATYC continued through the 1990s. In 1990, AMATYC hosted a national conference in Dallas, Texas, with over 1,000 attendees. AMATYC also began offering summer institutes to enhance professional development. Recognizing the high attrition and low success rates in developmental math, the AMATYC developmental math committee dedicated more time to improve the discipline. In 1994, AMATYC celebrated its twentieth anniversary with a major conference and celebration in Tulsa, OK (Blair, 1999).

AMATYC continued to soar as the decade progressed. The organization published its first major paper, *Crossroads in Mathematics*. The paper elaborated on standards as well as pedagogical approaches to assist students in community college courses below calculus level. In 1996, AMATYC went fully online with its own website. Individuals could access conference information, contact information, and an electronic copy of *Crossroads in Mathematics*. By 1999, AMATYC membership grew to 2,270 members. This included administrators, full and adjunct faculty, and even students. In addition to annual national conferences, AMATYC continued to hold various institutes to help the professional development of educators (Blair, 1999).

NADE (National Association for Developmental Education)

Developmental education continued to flourish in the 1990s. NADE continued to hold national conferences throughout the 1990s, and such conferences continued to exceed over 1,000 attendees (Boylan, 2016). In 1990, NADE conducted the first study for developmental education. In this study, researchers gathered data from over 5,000 students across 120 institutions. More specifically, researchers studied developmental education course structures, course outcomes, and instructional methods (Boylan, Bonham, Claxton, & Bliss, 1992).

In 1995, spearheaded by a committee of NADE veterans, *NADE Self-evaluation Guides* was released. This publication allowed developmental education programs to thoroughly evaluate themselves. Shortly thereafter, the NADE Certification program was established. This program provided recognition for developmental programs that demonstrated high quality education (Boylan, 2016).

In 1998 NADE collaborated with the U.S. Office of Education and the Harvard Graduate School of Education to present the first Harvard Symposium on Developmental Education. In this symposium, postsecondary educators worked together with public school educators and government officials to enhance work with at-risk students. In summation, within a 20-year span, NADE went from start-up to a major national educational organization.

Course Structure (1995)

The content within the courses, as well as the math gatekeeper courses themselves, at Sisco Community College, Habyan Community College, Telford Community College, Lester Community College, and Kilgus Community College did not change much from the 1980s to the 1990s.

> You can only shift around math content so much, *explained Professor Bell.* You need the arithmetic concepts to be successful in algebra. You need to know signed numbers and how to evaluate expressions for elementary algebra. You need to know factoring to do rational expressions. You need to know quadratic equations and roots and radicals plus all of algebra to do higher level math like trigonometry and pre-calculus, so what else can you do?
>
> I think our focus was more on how we could better teach these topics, *reflected Professor Mitchell.* How many ways can you arrange math topics?

Regarding other colleges, Griffin Community College kept their course structure together; however, the algebra 1 class moved into the developmental math department. Bordi Community College retained their course structure for their algebra and statistics courses; however, they split the initial arithmetic course into two separate courses:

Basic Concepts of Arithmetic I

This course focused on operations with whole numbers, fractions, and decimals.

Basic Concepts of Arithmetic II

This class reviewed whole numbers, fractions, and decimals. Emphasized in this class were also ratios and proportions, percentages, units of measurements, and operations with rational numbers.

Rationale for the Restructure

Our students just struggled so much with fractions and even whole numbers, so in 1992, we developed that additional course, *recalled Professor Milacki*. It was really the fractions. Students just struggled so much with fractions. We figured if we had one course which focused a lot on fractions and then review fractions in the next class, students would do better.

Professor Milacki added:

I think we were trying to emulate the K-12 curriculum somewhat. That is where the curriculum is cyclical. Students are exposed to fractions and other topics in third, fourth, fifth, sixth, seventh, and eighth grade. With more repetition, we figured it would help student success rates.

Did this improve student success?

We really weren't keeping track of student success rates in the early 1990s, but I don't think so. When students came to us deficient in basic skills, there just wasn't a whole lot we could do.

Professor Milacki added:

I guess it helped because the really low-end students could take the basic arithmetic 1 class, which was better for them and everyone else. When we had only one arithmetic class, those students would get lost early on and get only more lost as the quarter went on. I was spending so much time helping them, it took my attention away from the other students.

General Mathematics

The arithmetic, algebra, and statistics courses as well as the course content remained the same at Telford Community College. However, after completing the algebra sequence, students could enroll in general mathematics to gain credit for an associate degree as well as transfer credit. This class focused on data description, set theory, probability, logic, linear constraints, simple and compound interest problems, and annuities.

We wanted to develop a course for the non-math major. Actually, it was a course for the non-science or [non]engineering major as well. Basically, this was a math course, other than college algebra or statistics, that students could use to transfer, *recalled Professor Mitchell*. They still struggled in the class. Whether they took college algebra or statistics or general math, it really didn't make too much of a difference.

Instructional Modalities

Lecture Plus

Traditional face-to-face instruction continued to be the primary teaching modality throughout the 1990s. However, the participants asserted that they continued to employ interactive practices to engage their students. Professor Mesa expounded:

> My first year, I would explain a problem and then I would ask a question to the class, and I would hear the sound of one hand clapping. So, then I would answer the question myself.

It took an observation from Professor Mesa's mentor to note a major flaw in her teaching.

> My mentor said, "You're making a key mistake, which is asking a question to the entire group. When you do that, you'll rarely get an answer, and worst of all, you'll have no idea if they are getting it or not."

Professor Mesa began to look for ways to engage her students.

> When I started a problem, like solving a linear equation with fractions, I told the students, "We are going to solve this together", and before each step, I would call on a student to help with that particular step. If the student struggles, I would either ask if someone else could help or I would simply call on someone else. I could just tell it kept the students on their toes in class more, and it helped me understand if they were understanding.

Professor McDonald, a new instructor, found support from his colleagues regarding effective pedagogy.

> My department would hold these workshops once a year on Saturday mornings. We would have adjuncts there as well. People would take turns sharing teaching practices that worked. I found out pretty quickly that just lecturing wasn't very effective, so I was all ears.

Another faculty member gave him an interesting idea on how to engage students and quickly assess them as well.

> I started asking for hand signal responses. I would call on a student to give an answer in class, and then every student had to either give thumbs up if they agreed or thumbs down if they disagreed. Everyone had to participate. I also started using the hand signals when asking if students understood a concept or not. I know; I know, they could always BS me by giving me thumbs up that they understood something when they didn't, but I still think this helped the shyer or less confident students. They could give me a thumbs down if they really didn't understand something.

Phasing Out the Emporium Model

Previously, Lester Community College, Bordi Community College, and Kilgus Community College previously employed some variation of the emporium style for introductory math classes. However, the emporium model dissipated during the 1990s.

> I think we just realized students needed structure in their instruction, *recalled Professor Milacki.* Things were just more organized when everyone was on the same page. Also, by offering more traditional instruction, we could help students with their organizational skills. It helped that our class sizes were smaller. That [large class sizes] was the main reason I shifted to more of a self-paced model.

Lester Community College, which used the emporium model for its developmental math classes, shifted to more traditional instruction but for different reasons.

> We were trying to get more alignment with the math department. We were allowing students to work at their own pace and take tests whenever they were ready, but they [the students] were running into problems in the math department, because their policies were more strict. They weren't used to hard deadlines.

Professor Johnson explained the phase out of the initial emporium model at Kilgus Community College:

> It was phased out for a variety of reasons. The main reason was we didn't get the classroom [for the emporium model] anymore, and the Apple IIs went away. Also, the faculty, who were teaching in that method, gradually left.

Contextualization

Faculty also realized the importance of putting math into real-life context. Professor DeLeon elaborated:

> In traveling to AMATYC, I attended some interesting workshops about the importance of relating math to the real world. I tried harder to come up with word problems to make math more relatable to students' everyday lives.

However, Professor Johnson was not satisfied with traditional contextualized math problems.

> I just thought the problems were pretty dumb. A typical problem was "Brian pulls out two socks from his drawer. His drawer has seven black [socks], four brown, and eight blue. What is the probability that he gets a matching pair?" And I was like "who cares?" Usually, you match your socks before you put them in a drawer, and you turn on the light and you can see what color pants you're wearing to match them.

Professor Johnson provided another example.

> And I thought about the problems with nickels, dimes, and quarters. If you know the total change in your pocket, you probably know how many nickels, dimes, and quarters you have, right?

Professor Johnson wanted examples that were relatable to the modern world.

> It was the early 90s, and I thought we could explore more contemporary topics. I started creating problems that focused on the HIV rate and how likely or unlikely certain people or populations were to get it.

Professor Johnson added:

> I did those type of problems with a class that had a lot of nursing students. I also incorporated other medical questions in those classes.

Professor Morgan began to use more contextualized examples as well.

> I started gaging my students' interests and creating more examples based on those interests. For example, if I had a lot of football fans in one class, I would try and come up with probability questions focusing on football.

However, Professor Morgan noted a challenge with contextualization:

> I could come up with contextualized problems in arithmetic and statistics but not algebra. No way could I relate algebra to real life. It made answering their [the students] eternal question of "When am I ever going to use algebra in my life" very difficult.

The Development of Distance Learning

In the 1990s, distance learning grew rapidly in higher education, and community college math courses were no exception. The modality of distance learning met the need of a changing student population. Professor Ballard explained.

> It just seemed that students were getting busier. When I first started, I had a bunch of 18 and 19 years olds where school was their job. By the [19]90s, many of my students were working part-time or full-time. Some were married and even had children. Our administration was worried that we were going to lose students because of their schedules, so we were looking for alternate methods of instruction.

Networking helped Professor Ballard learn about distance learning.

> In the early [19]90s, there was a lot of discussion about distance learning at AMATYC and other conferences. I was just stumped about how to even do it. How do you even run a distance learning course? How do you help students learn math when they are not on campus in a classroom?

These were common questions at the time, but schools were finding answers. One common method was lecture through VHS tapes accompanied by some sort of guided practice for students to complete the course. Professor Ballard expounded:

> We started small in 1990. We decided to create a distance learning version of our algebra 1 class. That was the class with the highest enrollment, so we thought it would serve the most students. My colleagues and I created short lectures for each section. We filmed them with the help of the audio-visual department at Sisco.

Professor Ballard explained the process:

> In hindsight, it was a lot of work. We filmed in front of a camera, so we would have to wear certain clothes and they even had to put make up on us. I had a receding hairline, so they had to polish my head! Anyway, the folks in the audio-visual department packed the VHS tapes in the order we asked them to.

I asked Professor Ballard how students gained access to these materials.

> When they registered for the course, they had to buy the package of tapes. We had that worked out with the bookstore. They still had to come to campus to buy stuff from the bookstore.

Lester Community College utilized VHS tapes as well for the distance learning students.
Professor Fenimore explained:

> I also created a workbook for students to follow along and take notes and work problems as they went through each module [section or unit] on each tape.

Griffin Community College began employing a similar model in the early 1990s. Professor Thurmond recalled:

> Students watched the tapes and completed practice problems on worksheets.

However, Professor Thurmond noted a difference between 1990s and present-day distance learning.

> It wasn't like today where you had students all over the country or even all over the world taking your classes, and you never spoke with them. These were local students, and they came to campus. Many of them would work on their problems in the math tutorial center and come to my office hours. They just appreciated a more flexible schedule because of their work or family obligations.

Other schools employed cable television to reach their distance learning students. This was the case at Kilgus Community College. Professor Johnson explained:

> I would be live on a specific channel and teach the lesson. Students could watch me on television. I would even have three or four students come into a classroom and watch me, but most of the time, they would record it [the lesson] on their VCR and then play it back when they got home.

Bordi Community College began utilizing cable television for distance learning as well.

> I was really nervous before I started teaching, because I had no idea what I was doing, *shared Professor Milacki*. But we had brilliant people in the audio-visual and IT departments because they made it work. I just showed up and taught my lessons.

I asked the faculty how they assessed student-learning in these early days of distance learning.

> Students would come into the math learning center to take proctored paper and pencil exams. We had deadlines, but there was some flexibility, recalled Professor Fenimore.
>
> The deadlines for the exams were in the syllabus. Students would have to come into the math tutorial center to take the exam with a proctor, reflected Professor Milacki.

It is worthy of note that this period was prior to the wide use of email communication. Furthermore, this was before faculty could post students' grades to some sort of online interface. How did faculty communicate with students regarding their progress or answer students' questions?

> I would communicate with students on the phone. I would let them know their test scores and how they were doing in the class, reflected Professor Fenimore.
>
> Like I said, I still had a lot of students who would come in during office hours and ask me questions. I also made it a point to stop by the math tutorial center sometimes and I would see some of my DL [distance learning] students, *said Professor Milacki*. But regarding exams, I would let them know by phone.
>
> I would call students to let them know their grades, but I also used to post the grades on my office door. You know, I would only have the students identifiable by their social security number. That was back when you could that. Now, that would violate a whole lot of privacy issues, stated Professor Bell.

By the 1990s, it was apparent that students struggled in math. Did students struggle in the distance learning format in this period? Professor Fenimore stated:

> These were very good independent learners. They seemed to know that they needed to stay on top of things and learn independently when they signed up for the class.

Professor Thurmond concurred:

> In the early days, my DL [distance learning] students were my best students. They just knew what needed to be done, and they did it. I'd say I got the very high-end students who enrolled for my DL classes. They probably just needed a little brush-up.

Professor Mesa added:

> I would talk to these [distance learning] students, and they were just so matter of fact, so driven.

Professor McDonald made a good point:

> The success rates in my distance learning courses were higher than my regular classes, no question, but it's hard to compare because the enrollment in my distance learning courses was so much lower. You are comparing one large group to a tiny group, so I feel like it's comparing apples to oranges.

As the 1990s progressed, so did technology. This allowed some institutions to upgrade their method of teaching distance learning.

> I think it was around 1996 or 1997 we transitioned from VHS tapes to CD-ROM. It was just better quality. It was a lot of work to redo all the lectures, but it's what you had to do. We wanted to provide the best quality of instruction to our students, reflected Professor Mesa.
>
> It seemed like toward the late 1990s, most colleges were transitioning away from VHS tapes and instruction via cable television was a thing of the past, *recalled Professor Milacki*. We were looking for better ways to help students. A colleague attended AMATYC and learned about math on CDs. Again, our IT department were saints and geniuses. They helped our faculty design those lessons.

By the mid- to late 1990s, most institutions of higher education had networked systems, and this led to an updated form of communication with distance learning students.

> I don't remember the name of the program, because we've change so much over the years, but around 1997, our college adopted a system where we would record students' grades and post announcements and comments and whatnot. It was online, so students could access the information from anywhere, reflected Professor Bell.

This, however, was a difficult transition for some.

> In 1997, we started using the Edline system for online grading, *recalled Professor Ballard*. It was supposed to make everything easier. No more calling students; we could meet with students much less. Everything, grades, announcements, syllabi, would all be online. It was a nightmare. Of all the technological advancements, this had to be the most difficult. It was a hard system to learn, and we had so many technical problems. Students wanted to know their grades, and it was frustrating when I couldn't get them online because of the stupid Edline.

Professor Ballard noted another issue.

> I felt like online grading took the personality out of those classes. I missed talking with students. I just felt like by talking with them, I could help them more. With online grading, there were students whom I never talked to once during the quarter. That was weird.

Email communication became part of the higher education culture in the mid- to late 1990s. Due to the nature of a distance learning course, faculty did not have a choice but to adopt email. Professor Morgan reflected:

> We started emailing as a form of communication in 1996, and I didn't like it one bit. Email took the place of talking with students. My distance learning students used to come to my office, or I would meet them in the math tutorial, but now we were just writing to each other.

Professor DeLeon concurred:

> I couldn't explain math using email. It drove me crazy. I resisted it at first. I told students, "Let's just talk on the phone."

Professor Milacki touched on the changing nature of distance learning.

> It [distance learning] changed toward the late [19]90s. Enrollment in those classes went through the roof. We were getting more students who weren't coming to campus and truly completing the class from afar.

Professor Ballard added:

> I got the feeling in the early [19]90s, distance learning was a well-kept secret, and only the high-level students registered for those classes. I guess word got out and more students started to take these classes. By the late [19]90s, I didn't have as good a students in those classes. I had much more unprepared students.

This aligned with Professor Mesa's experience with distance learning.

> In 1994, I had the best students in those classes. They just knew what they needed to do and took off. By 1999, I was getting students who didn't even know what they were supposed to do. I had to hold their hands in showing them where they got the materials, when assignments were due and everything else.

Continued Efforts to Improve Student Success and Development

Faculty continued to search for innovative ways to help their students as well as develop existing methods.

The Continuing Development of Math Tutorials

Math tutorials and learning centers were major developments in the 1980s. These facilities continued to grow in the 1990s.

> With the help of the audio-visual department, I made more videos to help students. We had a great staff, and more and more students came to the math learning center, recalled Professor Fenimore.

Our math learning center had started off as a mom and pop shop, and by the mid-1990s, it had grown into a pretty busy place. We had student-tutors; we had professional tutors, and they had bachelor's degrees. We had two coordinators running the place. We had hundreds of students coming through our doors to get help, shared Professor Ballard.

We offered tutoring for our students, but by the [19]90s it became so much more than that. During the semester, we offered study skills workshops for students. At the end of the semester, we would offer final exam review workshops for students. These were workshops where a tutor would review for the final exam, reflected Professor Thurmond.

Our students raved about our math learning center. We had the perfect mix of caring people to help our students. What was really gratifying was we had students who would pass the introductory math courses and then become tutors themselves, recalled Professor Timlin.

Student Outreach

In the 1980s, faculty began working with academic counselors and this continued in the 1990s.

The counselors were so important to our work, especially with the developmental math students. When I started teaching in 1993, I had mostly developmental math classes. I thought teaching in college would be easy, but I was wrong, *explained Professor Mesa*. I had a lot of students who just weren't doing their work. They just seemed disinterested. Worst of all, I had a lot of behavioral issues. Tim, our department counselor, really helped me. He read some of the students the riot act about being a good student, and it helped.

I think it was around 1992, our department counselors would come around to all the developmental math classes during the first couple of weeks of school. They would introduce themselves and talk about student responsibility. I could always tell this was well-received by the students, reflected Professor DeLeon.

In the late 1990s, Lester Community College developed the math retention program to help students. Professor Wallace elaborated:

We had computerized practice exams for all our unit exams for elementary and intermediate algebra. We had two dedicated faculty members who offered study skills workshops.

Professor Wallace described the retention program's catch-up sessions:

If there was a student having trouble in elementary algebra, and got behind, they could attend a catch-up session. These sessions ran one or two weeks behind and were intended to get students caught up.

Like Griffin Community College in the 1970s, Lester Community College utilized first-day diagnostic exams.

These were in-class diagnostic exams that we gave students on the first day. It was a brief exam to make sure that students were placed into the right class. We identified students who may need to move down a math class or just students who may need review. If they needed review, we had prescription pads. For example, if the instructor had a student in intermediate algebra, and the student didn't know how to solve first degree equations, the instructor gave them a prescription pad and sent them to the math lab. And in the math lab there was a unit on first-degree equations for them to get help.

In this program, faculty took a proactive approach toward potential student struggles.

We had a phone call program where we called hundreds of students before the quarter. We went through and looked at their records, their performance the previous quarter, or their diagnostic tests. We called the students who we anticipated struggling in the upcoming course for whatever reason. We offered them private tutoring sessions, or we encouraged them to take a previous course again to gain a better understanding of the prerequisites.

As chair of the math department, Professor Milacki made it a priority to increase student outreach.

I got together with my full-time faculty, and we discussed ways we could be more proactive with our students. We decided to use three-week progress reports. Every three weeks, instructors would give students a report that listed their current grade with some comments. So, we worked together to create a standardized form.

Professor Milacki continued:

I'm not sure if it increased success rates a whole lot, but we just wanted students to be aware. I was so tired of it getting to the end of the quarter, and students had no idea why they were failing.

Women in Transition

Throughout the 1990s, Griffin and Habyan Community Colleges employed an outreach program focused on women who were recently divorced and looking to improve their quality of life. The program, Women in Transition (WIT), started in the 1980s but grew immensely in the 1990s. In Chapter 3, Professor Thurmond discussed the increased number of divorced women returning to school and taking advantage of open access education. Consequently, this program was very timely.

The program [WIT] has several components. First of all, it is state supported, so it is free. WIT helps these women register for college classes, but it also offers classes in helping them to live independently, look for jobs, and some classes in basic math, reading and writing. I coordinated the math part of the program at Griffin, reflected Professor Thurmond.

Professor Thurmond shared the logistics of the math instructional portion of WIT.

> It was before they actually took a math class. It took place in our math tutorial center. The women took a [math] diagnostic test, and we designed an individual program for them. For some women, it was arithmetic; others it was basic algebra, for some, it was a mix of the two. The women would watch our video tapes and complete worksheets. More importantly, they would get one-one-one help from one of our tutors. That was the most rewarding.

Professor Morgan was involved with the WIT program at Habyan Community College.

> We were serving a group who had been previously marginalized or not even recognized. Many of these women got married when they were really young. They hadn't worked or developed any kind of skills. Many were even in abusive relationships. It was really scary for them to now be out in the world on their own. Add to all of that, they were terrified of taking math.

Professor Morgan added:

> I had several women who would hug me when they finished with the math skills class. They came to us with low self-esteem and felt worthless. They didn't think they could do anything, let alone fractions, decimals, percents, and signed numbers.

The WIT program still exists in the present day:

> I'm retired now but I still go in and help out and talk to the women, *shared Professor Thurmond*. I was fortunate to be able to come from a good family and receive a good education that helped me have a good career. There are still many women who are not as fortunate and need this.

Calculator Policies of the 1990s

Community college math courses began using calculators in the 1980s. This continued in the 1990s; however, their policies and usage became more complicated.

> Our policy was always no calculators for the arithmetic or the basic algebra courses, and we always agreed on that, but by around 1995 or 1996, that started to change. Students were coming from high school and were shocked and upset that we didn't allow the use of the calculator. They couldn't believe they had to learn fractions and signed numbers without the calculator. Then, the faculty started disagreeing with each other over whether we should allow the calculator, reported Professor Milacki.

Professor DeLeon concurred:

> It was really the first time we had strong disagreements in our depart-
> ment. We had some faculty, like me and I was chair at the time, who
> thought that students should not be using the calculator for such basic
> stuff like fractions, decimals, percents, and rational numbers. Other fac-
> ulty members thought we should use the calculators since students were
> already coming expecting to use the calculator. We kept the non-calculator
> policy, but it was a struggle.

In 1998, the math department opted to allow the use of four-function calculators at Griffin Community College (GCC) for their algebra classes. This created more pressure for the developmental math department at GCC.

> The math department decided to go all calculator because they were
> sick and tired of students not knowing how to compute fractions, and
> decimals and signed numbers by hand. It only intensified the arguments
> within our department. Some of us believed that a big reason students
> tested into developmental math was that they lacked number sense, and
> we felt being able to compute arithmetic concepts by hand would help
> develop that number sense, *shared Professor DeLeon*. Others were like the
> faculty in the math department. They were just sick of having students
> learn everything by hand. Again, their argument was that more and
> more students were coming out of high school as calculator dependent,
> so why should we change the world?

The developmental math department at Telford Community College was fac-
ing a similar divide regarding calculator usage. Professor Mitchell reflected:

> In 20 years of teaching at the community college level, this [the calcula-
> tor debate] was the first time our faculty were really divided over an
> issue, and that division was not friendly. I was part of the "students in
> developmental math should not be using calculator" group. I always
> asked, and no one would answer, "What if the calculator breaks or
> runs out of batteries? How will the student add two fractions then?"
> But we had faculty who not only believed students should use the
> basic [four-function] calculator but also the scientific calculator. It was
> a mess. It was the first time we had department meetings with people
> yelling at each other.

Professor Mitchell recalled another issue driving calculator usage.

> We had academic counselors and for the first time, administrators, ask-
> ing why we are not allowing the use of the calculators. Students were
> complaining that they could use the calculator in high school, and why
> were we being so mean? AMATYC was also pushing the use of the cal-
> culator, so that didn't help our argument. It was the first time I actually
> felt my expertise being challenged. I never realized it, but administrators
> always treated us as the experts in our field. If they had a question about
> why we were doing something, our response was good enough. This
> was the first sign that things were changing.

The scientific calculators were increasingly used in the college algebra, statistics, and even intermediate algebra classes without much controversy.

> I think they [the scientific calculators] took some getting used to, but many of us were appreciative of how much computation was eliminated. Also, I think in the past, even in classes like stats, trig, and college algebra we would try to stay away from messier problems because we didn't have the calculator, and we wanted students to have enough time on exams, but the calculator solved that issue, recalled Professor Bell.
>
> There were professors, like me, who had been teaching since the 1970s who thought the calculator, especially the scientific calculator, would dumb down math, *explained Professor Morgan*. But it just allowed us to give different types of problems, even harder problems. Also, the scientific calculator was a tool, and we were teaching them how to use that tool.

As technology continued to evolve, higher education was introduced to yet another type of calculator, the graphing calculator. In 1985, Casio introduced the first graphing calculator. The graphing calculator had the same capabilities as the scientific calculator; however, graphing calculators could perform the same computations, such as simplifying radicals, with less steps. Moreover, these calculators could also graph various functions with the capability to adjust the windows. Sisco Community College implemented the TI-83 graphing calculator for their introduction to statistics class in 1996.

> We didn't come to the decision lightly or quickly. First there was the cost. I want to say those suckers [the calculators] were about $125. Those were more expensive than the textbooks. Then, there was the learning curve. Some of us were worried if we could learn the graphing calculators well enough to teach the course well, shared Professor Ballard.

Fortunately, for Professor Ballard, there were faculty who were willing to help other faculty members learn the graphing calculator.

> That was the great thing about Sisco. It seemed like we always had faculty willing to help everyone out. We had these two guys, one was part-time; the other was full-time, and they had a lot of experience with the graphing calculator. They had both taught at other schools that used them.

For others, the introduction of the graphing calculator was not a smooth transition. Professor Timlin elaborated:

> We tried the TI-83 in 1997 for our college algebra classes. It was a quick and not so well thought-out decision to be honest. Some of our faculty attended AMATYC and were convinced that this would help our students. To be honest, I was tired of having students graph exponential and logarithmic functions by hand, so I was all right with it.

Professor Timlin struggled with this new technology.

> I'll be honest. I didn't prepare as much as I should have. I learned the scientific calculator pretty easily, so I thought this wouldn't be too bad. Wrong! I was struggling with the basic syntax of getting through a

problem. Worst of all, even if I got through a problem, if a student had a logistical or syntax question [using the calculator], I had a lot of difficulty trouble shooting a problem. It got to the point where I started using the calculator less and less in class, and the students were not happy about that. I can't blame them since they paid an arm and a leg for it.

Professor Timlin had a solution.

I went to my chair, and I told her I was not using the TI-83 the next semester. Fortunately for me, I wasn't the only one who struggled. So, using the graphing calculator became optional for the instructor. You could require it for your class if you wanted.

The optional use of the graphing calculator was the case at Kilgus Community College as well.

There was a lot of controversy among the mathematics community and big fights in my department, *recalled Professor Johnson.* Some faculty wanted it [the graphing calculator] and some didn't, so we listed the courses that required the graphing calculator, so students would know if they had to pay the $100 or so. What made it more complicated was we had classes all the way down to elementary algebra using the graphing calculator.

The Student Population Ages and Changes

The participants articulated that by the late 1990s, their student population looked very different compared to the 1970s.

It started happening in the [19]80s, but we had much more non-traditional students. For the first time, I had an 18-year old and a 65-year old in the same classroom, explained Professor Morgan.

A class during the day still had mostly 18 and 19-year olds, but an evening class had mostly older students. Even in a day class, you could get older students, recalled Professor Bell.

Why was there a shift in age demographics?

It's like I said for the students in the 1980s, you just needed more education, and that was even more the case by the 1990s. I would say by the 1990s, needing a college degree was like needing a high school degree in the 1970s. In my conversations with older students, they were talking about how they simply wanted a better life or some had jobs and wanted a promotion, but they needed a college degree, shared Professor DeLeon.

We also became the college of second and third chances, *recalled Professor Ballard.* Students who dropped out of college five or ten years prior were coming back.

The faculty also noticed a shift in the overall demographics of their students in the 1990s. Professor Morgan noted:

> When I started teaching, we had a predominately White student population. It seemed like by the 1990s, we had more Blacks and Latinos. We were getting more inner-city students. I think they [the inner-city students] were being told they should just go to community college after high school. I don't think that was the case before.

Professor Thurmond noticed this change as well and how it impacted the student population.

> We always had behavioral issues in the developmental math classes. Students were unmotivated and disinterested, but now it seemed like the behavioral issues were worse. We had students who didn't understand how to behave in a college class. It almost seemed like gang behavior at times. In 1999, I actually had a student curse me out in front of my class over a grade for the first time.

Professor Milacki concurred:

> We always got students who didn't belong in college, but this intensified in the 1990s. I know this sounds bad, but I felt like we became a referral place for a lot of the inner-city students. It's like someone there [the inner-city schools] was telling them to come to Bordi [Community College]; take classes and you'll get financial aid.
>
> I asked the participants if this change in the students' demographics impacted their preparedness level regarding mathematics.
>
> No, not really. Students struggled with the same concepts in the 1990s that they struggled with when I started in the 1970s. They didn't understand fractions, order of operations, signed numbers, equations with fractions. Just more behavioral issues, responded Professor Ballard.
>
> The calculator was creating some weird issues. We had students who actually completed algebra in high school, but they were testing into much lower algebra or even arithmetic. The problem was they were calculator dependent, but we didn't allow the calculator on our placement test, *shared Professor McDonald.* There was resentment on their part. Actually, there was resistance. They didn't understand why they needed to learn fractions and signed numbers in college if they already had them in middle and high school.

Professor Mitchell agreed:

> The students still struggled with fractions, decimals, and rational numbers like they always did, but I felt like they were harder to teach when they came to us calculator dependent and couldn't understand why they couldn't use the calculator.

The Rise of ESL Students

Included in the changing student population was the increase in the number of English as a Second Language (ESL) students in community college math courses. Professor Morgan elaborated:

> This was just reflective of the 1990s. We had more and more students coming to college who hadn't before, and that included more ESL students. Here is what people don't understand. Yes, community colleges became open access in the 1960s, but it took time for more and more students to realize they could go to college.

Professor DeLeon added:

> It's like I said before. The job market got so much more demanding by the [19]90s. It wasn't enough just to have a high school diploma. People needed more education and more skills, and that included ESL students.

The Challenges for ESL Students

Some ESL students faced immense challenges when they attempted community college math. Professor Mesa explained:

> We were getting students who could barely speak English placing into developmental math. It was a diverse group. We had a lot of Spanish-speaking students. We had some Indian students and some Arab students. You could see they were confused and scared.

Professor Timlin added:

> It wasn't that they [the ESL students] couldn't do math. They had such language barriers. Their English was so deficient that they couldn't follow basic instruction. They had trouble just understanding what they needed to do for class.

Professor Morgan provided additional clarification:

> Math is a universal language. People all over the world do fractions, decimals, and solve equations. The problem is the student needs to be able to understand instruction. The students need to be able to communicate their questions. When there are major language barriers, the students can't do that.

An additional challenge for some ESL students was that they learned math differently in their native country, and this created confusion when attempting community college math in America. Professor Mesa clarified:

> I had some ESL students who would look so confused during class, and when I would work with them individually, they would tell me how I was explaining things in such a different way than they were used to, and it was confusing to them.

Professor Bell mentioned that this difference in learning was more for the developmental math students:

> The ESL students in my statistics or my calculus classes had no problem. It didn't matter where they learned math; they caught on fast. It was the weaker students. Their math foundation was so weak that seeing it [math] explained in a different way or looking different just blew them away.

I asked the participants if their respective community colleges assisted ESL students as they transitioned to community college.

> There were ESL classes for students, but I think those classes benefitted students who already knew a good amount of English, but we really didn't have sufficient services for students who were really deficient in English, recalled Professor Mitchell.
>
> I felt like it was our responsibility to help the students who really struggled with English, and that wasn't fair. That requires special training, and I didn't have that, asserted Professor Bell.

Teacher Preparatory Classes

Throughout the twentieth century, undergraduate colleges and universities offered teacher preparatory programs for pre-service teachers. In such programs, students studied methods of teaching disciplines such as science, social studies, reading, and of course, math. By the late 1990s, many community colleges were offering teacher preparatory classes in math for students who intended to pursue a bachelor's degree in elementary education.

Community colleges implemented these courses for a common reason. Professor Timlin explained:

> Since the nearby colleges and universities required these [teacher preparatory] classes, students could take them with us and then transfer. It helped our enrollment.

The faculty mentioned that they aligned the curriculum for their math teacher preparatory classes with the curriculum of the local four-year colleges and universities to ensure a seamless transition for the students. The content in these classes includes concepts that encourage critical thinking such as number theory, set theory, probability as well as data description. Students are also introduced to manipulatives such as base ten blocks for teaching whole numbers and decimals as well as Cuisenaire rods for teaching fractions.

All the faculty participants stated that intermediate algebra was, and remains, a prerequisite for the math teacher preparatory classes.

> Students need a deep conceptual understanding of arithmetic and algebra when they come into these classes, *asserted Professor Ballard*. They need to apply their math skills. I've long thought that college algebra should be a prerequisite, so the students had an even stronger background.

Professor Bell concurred:

> Students need a strong math background for the teacher prep classes, but many of them don't have it. I have always needed to teach the basics.

In his book *How to Succeed in College Mathematics*, Richard Dahlke (2008) asserted that math preparatory classes are difficult to teach, and it can take time and practice for faculty to become skilled at teaching these classes. Professor Milacki agreed:

> It's a balancing act. Students often enter these classes with some weak skills. There is basic content and basic algorithms they don't remember. So, you need to be able to provide quality instruction, help them become familiar with the manipulatives and help them to think critically with the abstract number theory and set theory problems.

Shifting Relations between Faculty and Administrators

Throughout the 1970s and 1980s, the faculty generally reported positive relationships with their administrators. Their administrators were supportive and respected the faculty as the experts in their field. For some, this continued in the 1990s.

> I remember positive interactions with my dean and even president, *recalled Professor Mesa*. To be honest, they just kind of stayed out of our way and let us do our jobs.
>
> Our dean was a huge help in the expansion of our academic support center. She listened to our needs, fought for funding, and was very appreciate of what we were doing, *recalled Professor Thurmond*. I was sorry when she retired in 1998.
>
> I remember how supportive my dean and my president were about us venturing into distance learning. We were one of the first on campus to do that [teaching via distance learning] and they gave us their complete support, *shared Professor Milacki*. We decided to try distance learning as a faculty. The dean gave us his support and even acknowledged the work we did at an annual meeting. In hindsight, I never realized how good we had it, or how much I would miss it.

Some respondents noted a change. Professor Morgan clarified:

> I had the same dean, provost and president from the time I started to the 1990s. I want to say my dean retired in 1995, and the president and provost in 1996. I couldn't put my finger on it at the time, but something was different about the new people [dean, provost, and president]. They seemed more distant. Our former president, Dr. Hentgen, knew us by name; he knew my family; he would come and sit with my

colleagues and I at lunch. This new president, Dr. Key, seemed nice, but he just wasn't as involved. He didn't make the effort to get to know the faculty.

Professor Morgan continued.

Our new dean, Dr. Davis, was different as well. She would ask us questions about our teaching, but it was different than my previous dean, Dr. Smith. Dr. Smith was genuinely interested and excited in what we did in the classroom and with our students. When Dr. Davis would ask us questions, I almost felt like I was answering questions at a deposition.

While Professor Morgan had difficulty putting her finger on the change, Professor Ballard did not.

During my time [1970s–1990s] at Sisco [Community College], I had two presidents and two deans, and both were great. Did we always agree on everything? No? But they respected us as the experts in our discipline. I didn't even appreciate that at the time.

Professor Ballard continued:

In 1998, we got a new dean. I called him "Preppy Harvard." He had a business degree and a doctorate in leadership, and I could tell this guy was was different right away. He just seemed to talk down to us. I get that he was our supervisor, but my previous deans treated us more like colleagues. Anyway, in 1999, he visited a math department meeting and was asking questions about our curriculum and why we were teaching what we were teaching. It just didn't feel right. I could tell this guy didn't like us. That was fine, because I didn't like him.

Professor Mitchell recalled a specific issue with her new dean.

We got a new dean in 1995, and all was well at first, but he called a meeting with us in early 1997 to ask us about the calculators and why we weren't allowing our developmental math students to use a calculator. He was citing the fact the students in high schools use calculators and we didn't. He didn't direct us to use the calculator, but it was the first time an administrator was meddling in what we were doing. I didn't like it, and it didn't make the arguments we were already having about the calculator, in our department, any easier.

Math Wars Brewing Below

In the 1990s, the primary instructional method for community college math was face-to-face. While the term, "lecture" was still employed, faculty were employing more variety (e.g., collaborative learning, study skills) in their pedagogy. However, variations on pedagogy were generally accepted within higher education. This was not the case in K-12 mathematics.

The traditional modality of instruction in K-12 was lecture-based with drill and practice. While there was a progressive movement, in education, earlier in the twentieth century, these views were largely ignored over time. That changed in 1989 when the National Council of Teachers of Mathematics (NCTM) published *Curriculum and Evaluation Standards for School Mathematics.* This document contained a set of standards judging the quality of math pedagogy. In general, NCTM recommended a shift from lecture and drill and practice to more of an inquiry- based approach where students would learn collaboratively with the teacher serving as a facilitator. NCTM also pushed for project-based learning and assessments as opposed to traditional homework and paper and pencil exams. With emerging technology, NCTM lobbied for more calculator usage in the primary and secondary education classrooms (National Council of Teachers of Mathematics, 1989). This reform was based on data that showed that American students struggled with problem solving skills and abstract concepts in math (Schoenfeld, 2004).

NCTM's push for radical change in math education created major debates within the discipline. Critics of this reform argued that students required traditional instruction before engaging in more abstract concepts. Additionally, these critics expressed concerns that students would become calculator dependent, and this would hinder their development of number sense. There was also the time factor. Students could complete traditional homework in a much shorter time than completing a group project. These "math wars" were simmering in K-12 the 1990s but would eventually gravitate to higher education in the future.

Summary

In the 1990s, the primary method of instruction was face-to-face lecture. However, faculty continued to engage students. The emporium model of the 1970s and 1980s were phased out. Additionally, technology continued to impact community college math. Some faculty acclimated to the scientific calculators while others attempted the novice graphing calculators. Distance learning made its way into community college math and evolved with the technology.

Faculty continued to work on student development and searched for additional methods to help struggling students. Math learning centers and tutorials grew and were well-received by students. Additionally, faculty developed outreach programs to assist students.

The discipline of community college math was about to move into the twenty-first century. On the surface, the discipline seemed to be soaring. Faculty continued to create and develop innovative ways to reach their students, and the advancement of technology could only help this. However, there were ominous developments in the 1990s as well. Calculator debates divided faculty into factions. The student population became more challenging. An administration that seemed less appreciative and more critical created tension. Some faculty felt that they were no longer treated as experts in their field. This, however, would be only part of what led to a tumultuous subsequent decade.

References

Blair, R. M. (1999). *The History of AMATYC: 1974–1999.* Memphis: American Mathematical Association of Two-Year Colleges. https://cdn.ymaws.com/amatyc.org/resource/resmgr/history/amatychistory.pdf.

Boylan, H. (2016). History of the national association for developmental education: 40 years of service to the field. https://thenoss.org/resources/Documents/History/History%20of%20the%20National%20Association%20for%20Developmental%20Education.pdf.

Boylan, H., Bonham, B., Claxton, C., & Bliss, L. (1992, November). *The State of the Art in Developmental Education: Report of a National Study. Paper Presented at the First National Conference on Research in Developmental Education,* Charlotte, NC.

Dahlke, R. (2008). *How to Succeed in College Mathematics: A Guide for the College Mathematics Student* (1st ed.). Plymouth, MI: BergWay Publishing.

National Council of Teachers of Mathematics (NCTM) (1989). *Curriculum and Evaluation Standards for School Mathematics.* Reston, VA: National Council of Teachers of Mathematics.

Schoenfeld, A. H. (2004). The math wars. *Educational Policy, 18*(1), 253–286. doi: 10.1177/0895904803260042.

5

The Aughts (2000–2009):
A Time of Reform and Turbulence

What Happened?

While the participants faced challenges, their experiences teaching community college math in 1970s, 1980s, and 1990s were very good. However, below is some feedback regarding their teaching experiences in aughts (2000–2009):

> It was a nightmare. I never felt so micromanaged in my entire life. Teaching during this time was an insult to me as a professional, said Professor Thurmond.
>
> My administration made my life miserable. I couldn't teach math effectively. How could I when I was micromanaged so much and couldn't do my job, said Professor Mesa.
>
> It was so hard to teach my students, because we kept having to redesign the curriculum. How can that work? Inquired Professor Mitchell.
>
> I guess I took for granted all the pedagogical freedom we had and how respected we were [in the 1970s, 1980s, and 1990s], because it all went away, proclaimed Professor Milacki,
>
> In 2009, I decided to retire, and I told Preppy Harvard [Professor Ballard's dean] that he was the worst administrator I ever worked for, and he turned our once fine department into a chaotic mess, said Professor Ballard.

What soured the faculty experiences this much in their discipline? Was it changes in the administration? Were there changes in the student population? How could the dynamics of their jobs change so much to instigate such negativity? Those questions, and more, will be explored in this chapter.

Returning Participants

In Chapter 5 Professors Ballard (Sisco Community College), Mitchell (Telford Community College), Milacki, (Bordi Community College), Morgan (Habyan Community College), DeLeon (Griffin Community College), Fenimore (Lester

DOI: 10.1201/9781003287254-5

Community College), Wallace (Lester Community College), Thurmond (Griffin Community College), Bell (Habyan Community College), Johnson (Kilgus Community College), McDonald, (Griffin Community College), Timlin (Griffin Community College), and Mesa (Telford Community College) shared their experiences.

Additional Participants

This chapter introduces five new participants.

Professor Sutcliffe

Professor Sutcliffe began teaching developmental math in 2003 at Griffin Community College (GCC). At that time, the developmental math and math departments were separate departments. This changed in 2016 when the two departments merged. She has a bachelor's degree in elementary education and a master's in education.

Professor Moyer

Professor Moyer began teaching developmental math and college-level at Telford Community College in 2004. He has a bachelor's and a master's degree in mathematics.

Professor Trombley

Professor Trombley began her career in 2005 at Sisco Community College (SCC). She has a bachelor's degree in math and a master's degree in math education with 18 graduate hours in pure math.

Professor Lopez

Professor Lopez started teaching at Bordi Community College in 2006. She has a bachelor's and a master's degree in math. Professor Lopez became chair of the math department in 2015.

Professor Mussina

Professor Mussina started teaching at Bordi Community College in 2004. He has taught both developmental and college-level math. Professor Mussina has a bachelor's degree in math and a master's degree in math education with 18 hours of graduate level math.

Developmental Math Gets Bombarded with Statistics

During the aughts, community college math faculty became besieged with statistics. These statistics focused specifically on developmental math, and such statistics were abysmal. Again, during the 1970s and 1980s, institutions did not focus on student success rates in mathematics. However, this started to change in the 1990s.

The first set of data that raised eyebrows regarding community college math was the high number of students who tested into developmental math via the placement test. Community colleges across the country witnessed this. By 1998, 91% of the incoming students in the Maricopa and Pima Community College districts placed into developmental math (Puyear, 1998). In the early 2000s, 61% of students who entered the Virginia Community College system had to enroll in a developmental math course (Curtis, 2002). This was even more so the case for the Colorado Community College system where 83% of the entering student population tested into developmental math. In 2007, Biswas reported that 80% of students who entered Housatonic Community College, in Connecticut, enrolled in a developmental math class. Overall, in 2006, Noel-Levitz stated that nationally 75% of all incoming community college students enrolled in a developmental math course.

The next set of data to assault community college math focused on the student success rates. More specifically, at what rate were students completing their introductory math classes? Several community colleges reported students struggling in these courses. The Virginia Community College system conveyed that as little as 29% of students were successful in the lowest-level developmental math course (Waycaster, 2001). In 2006, the Indiana Community College system reported that only 53% of students were successful in developmental math courses. Specifically, the developmental math success rates in the Indianapolis Community Colleges were a trifling 29% (Office of Institutional Research and Planning, 2007). The student struggles extended to New York City where in 2007, developmental math success rates were as low as 36% (Hinds, 2009). In 2008, a study of 107 California Community Colleges conveyed that 75% of students who endeavored a developmental math class were unsuccessful (Bahr, 2008).

As if the aforementioned data were not dreadful enough, community colleges were also seeing paltry persistence rates. That is, many students who attempted a developmental math class did not persist in their education. Only 9.6% of students who were unsuccessful in their first developmental math courses were still enrolled in their college 1 year later. In contrast, 66.4% of the students who passed their first developmental math courses were still enrolled (Boylan, 1997). Bahr (2008) reported that 81.5% of students who attempted a developmental math class did not stay in school or even transfer to another institution.

Success Rates Impact State Funding

Starting in the 1960s, open-access community colleges with low tuition were viewed as a vehicle to serve those who had been marginalized in education. States funded community colleges based on their full-time equivalent (FTE). If the institutions received healthy funding, college administrators were satisfied. However, this began to change toward the end of the twentieth century.

Between 1980 and 2000, the average amount of state funding allotted for community colleges and public 4-year schools decreased from 46% to 34% (American Council on Education, 2004). While all states endured cutbacks, by 2010, Alabama, South Dakota, New Mexico, South Carolina, Florida, Arizona, Virginia, Nevada, Utah, and West Virginia experienced the largest reductions in public higher Education (Hebel, 2010).

What was the cause for such reductions? Starting in the 1980s, two entities began drastically pulling at state budgets: Medicaid and escalating prison costs. Hefty surges in caseloads, escalating prescription costs, and waning support from the federal government led to high Medicaid costs. Regarding prison costs, state spending on corrections grew six times the rate of spending on higher education between 1985 and 2000 (Rizzo, 2006).

As other entities began tugging at state budgets, the states began examining community college success rates. This led state legislatures to the abysmal statistics related to developmental math. Legislatures began to determine that remediation was costly to American taxpayers (Apling 1993). More specifically, they argued that taxpayers should not have to pay for students to learn the same content in primary and secondary school and then again in higher education. On top of that, the paltry success rates indicated that students were not learning the material in higher education as well. In the early 2000s, state legislatures and higher education boards began to hold institutions more accountable for student success rates (Arendale, 2003). In doing that, states began to impose funding formulas on community colleges. Funding was no longer solely based on full-time equivalents. Consequently, schools needed to demonstrate better success rates to receive funding (Weisbrod et al. 2008).

Persistence Rates Lead to Course Restructure

The dismal statistics connected to developmental math coupled with the pressure from the states to increase student success rates led to increased oversight from community college administrators. In the late 1990s and early 2000s, developmental math departments came under scrutiny for persistence rates. More specifically, administrators noted that many students who passed the highest-level developmental math class were unsuccessful in the introductory credit-bearing course in the math department. This was especially the case for the schools where developmental math and college-level math were separate departments.

Lester Community College

At Lester Community College (LCC), during the aughts, the content in the introductory to statistics and college algebra classes stayed relatively the same. However, the developmental math courses saw many changes. Professor Fenimore elaborated:

> Our dean showed us data where only 28% of the students who passed our introduction to algebra class (developmental math) passed elementary algebra (math department). So, we were under a lot of pressure to get those success rates up.

Professor Fenimore continued:

> We got a lot of blame from the math department that we weren't challenging students enough. Students were having trouble with not just the content but the fast pacing of elementary algebra.

Their solution at LCC added more content to the introduction to algebra class.

> We beefed up introduction to algebra by adding in more on laws of exponents and factoring. We also put more geometry concepts in as well like cylinders and cones.

The faculty at LCC also noticed that students were struggling in their fundamentals of arithmetic class. In the early 2000s, the title was changed to basic mathematics. Professor Fenimore conveyed:

> The students were struggling so much in whole numbers. They were having trouble with fractions, but also, they were struggling with multiplication and division.

Professor Fenimore shared the solution:

> We created another arithmetic course (basic mathematics 1) that covered only whole numbers, fractions and decimals. That way we could help students with whole numbers and give them more exposure to fractions and decimals as well.

Telford Community College

Professor Mitchell recalled being blamed for persistence rates as well.

> We had all these meetings starting in the late [19]90s, and they continued into the 2000s. I think it was around 2000, we started having meetings with the math department about why our students who passed basic concepts [of algebra] were having so much trouble in elementary algebra. I think the persistence rate was around 30%.

Professor Mitchell did not recall these meetings fondly.

> I felt like we [the developmental math department] got blamed for everything. Students in elementary algebra had so much trouble keeping up with the class. They [the math department] acted like we weren't tough enough on the students.

Like the faculty at LCC, the faculty at Telford Community College (TCC) added more rigor to their highest level developmental math course.

> We started giving timed tests. The math department complained that we allowed too much time for exams. We started collecting written homework. Again, the math department faculty collected homework, and I guess a lot of our faculty didn't, so students weren't used to turning in homework. We also allowed students to retake one exam, and again, that wasn't allowed in the math department, so we stopped doing that.

The faculty at TCC added more content to the basic concepts of algebra course:

> We also added more content to our class. To help students adjust to more rigorous pacing, we added the laws of exponents and some factoring to our course. They already covered those topics in elementary algebra, but we thought if we at least introduced them [the students] to those topics in our course it would prepare them better. For the laws of exponents, we only covered the product, quotient, and power-to-power rule. Then, in elementary algebra, they would go more in depth with exponents like combined rules and negative exponents. Same with factoring we covered only factoring out the greatest common factor and factoring basic trinomials. We thought this would prepare them for the more rigorous factoring in elementary algebra. But in general, we figured it would help them adjust to the more rigorous pacing of the math department.

Griffin Community College

Professor DeLeon reported that the developmental math faculty at GCC came under fire for low persistence rates regarding students in their basic arithmetic course being unsuccessful in the algebra 1 course.

> I think only 32% of the students who passed our basic arithmetic courses passed the algebra 1 class.

When Professor Sutcliffe was hired, she had a standard meeting with the college president to sign the contract.

> The meeting only lasted about 10 minutes, but I remember him telling me that we had one of the lowest success rates in the country for students transitioning from developmental math to college-level math, and he added "The state notices that." I didn't think anything of it at the time, but man that must have been on his mind.

The math department at GCC complained that students exited developmental math with insufficient basic skills. Professor Thurmond expounded:

> They wanted us [developmental math] to emphasize fractions, decimals, and order of operations more in our classes.

Professor Timlin elaborated:

> We were using the four-function calculator in our algebra 1 class, but students still needed to able to understand the rules of fractions and order of operations. Even if they had them [fractions and order of operations] in the developmental math class, we had to reteach them.

GCC had the same solution as LCC.

> We split our arithmetic class into two separate classes. Our first, which we called basic arithmetic 1, focused on whole numbers, fractions, and decimals. This allowed us time to teach these concepts and then do a lot of drill and practice. In basic arithmetic 2, we covered those concepts again along with percents, ratios and proportions, measurements, and signed numbers. It just allowed for more time for them to learn these concepts, shared Professor DeLeon.

Course Restructure Reversed

Both TCC and LCC found their newly rigorous developmental math classes to be short-lived.

> We made our basic concepts (of algebra) class harder to try and prepare students for elementary algebra but then the individual course success rates for basic concepts dropped. Our dean freaked out because now less students were getting out of basic concepts, *recalled Professor Mitchell*. So, about a year after we redesigned our basic concepts of algebra class, our dean wanted us to remove some concepts from basic concepts of algebra. He said there was too much overlap, too many topics we were

covering in both classes. I tried to explain that this was to help prepare students for elementary algebra, you know to give students more practice, but he didn't listen.

In addition to being micromanaged by her dean, Professor Mitchell expressed frustration that the faculty in the math department changed their story:

> For years, they [the math department], complained that we were too easy on our students. Then, they started complaining that they wanted us to focus more on basic skills like rational numbers and evaluating expressions. They said we could leave the hard stuff, like laws of exponents and factoring, to them. Maybe they were just siding with the dean; I don't know. So, we removed laws of exponents and factoring from basic algebra. I was ticked off, because redesigning a course twice in a year is a pain.

Professor Fenimore noted a similar occurrence at LCC.

> Our success rates [in introduction to algebra] went down, and that caught the administration's attention. People in the math department started noting that there was a lot of overlap in topics between introduction to algebra and elementary algebra. So, we got pressure to go back.

Did adding rigor to the top developmental math class at least lead to higher persistence rates in the introductory course in the math department, which was the intended goal?

> I have no idea, *responded Professor Mitchell.* I kept asking, but they stopped keeping track of those persistence rates. All they cared about at that point was our individual course success rates. It was very frustrating.

National Initiatives to Improve Student Success in Math

In the 2000s, the dismal success and persistence rates in community college math were capturing national attention. Consequently, national initiatives emerged to increase student success rates. In 2003, several of the nation's community college experts were brought together by the Lumina Foundation, a private Indianapolis-based group. Their focus was on raising student success and graduation rates at community colleges. In conjunction with other private foundations such as the W.K. Kellogg Foundation, the Boston Foundation, and the Knowledge Works Foundation, the Lumina Foundation offered millions of dollars to community colleges across the country via grant-based funding. The initiative was named Achieving the Dream: Community Colleges Count (Ashburn, 2007).

Through the Achieving the Dream grant, community colleges examined their own success rates to determine ways to increase the success rates in gatekeeper courses. More specifically, the institutions developed their own initiatives to improve student success. Community colleges that received this grant were assigned coaches. These individuals worked with the colleges to help implement such initiatives.

Acceleration

By the 2000s, legislatures and administrators viewed developmental math as a barrier to student success. Consequently, colleges began to search for ways to accelerate students through their developmental math course sequences so that they would reach their college-level math courses in less time. This was based on the data that showed the more time students spent in developmental math, the less likely they were to persist in college.

By the late 2000s, the Bill and Melinda Gates Foundation became involved in developmental math. More specifically, the Gates Foundation pledged $110 million to develop innovative models to help students accelerate through developmental education (Ashburn, 2007). Consequently, many community colleges that implemented programs that focused on accelerating students through developmental math did so via funding through Achieving the Dream or grants through the Gates Foundation.

The Return of the Emporium Model

After dissipating in the 1990s, the emporium model returned and gained national attention when Virginia Polytechnic Institute and State University adopted the model later in the decade. Throughout the aughts, about 120 institutions adopted the emporium model through the Achieving the Dream initiative (Twigg, 2011).

The general design of the emporium model of the 2000s was similar in format to emporium models of the twentieth century. Students worked independently in a lab setting and then attempted exams when completing a unit or module. However, by the 2000s, mathematics software had become interactive and networked. More specifically, through programs such as MyMath Lab (later renamed MyLab Math), Hawkes Learning, and ALEKS, students were able to obtain instant feedback after completing a math problem, and their instructors were able to remotely access their work. Such software programs allowed students to move through material they understood and focus more on the content in which they struggled, and therefore, they could accelerate through their developmental math courses. By the end of the decade, schools such as Cleveland State Community College in Tennessee were reporting higher success rates in developmental math courses that employed the emporium model (Squires et al., 2009).

As part of the Achieving the Dream initiative, Bordi Community College (BCC), brought back the emporium model. Professor Milacki recalled:

> It was 2005, and a few of us were called into the dean's office to discuss this new initiative from this grant they just received. He [the dean] was telling us about the emporium model and how great it was. I even said out loud, "So, you want us to go back to what I was doing 20 years ago." He had no idea what I was talking about. He just kept telling us how it would get students through dev[elopmental] math and how other schools were doing it. Since I was the chair at the time, I was expected to implement it. It also kinda bugged me that as faculty we had no say in the matter. It was just administrators making decisions about how we teach our classes.

Professor Milacki discussed the creation of the updated emporium model.

> It wasn't that hard to put together. We had been using MyMath Lab as an optional tool, so some of us were familiar with it. I also relied on my experience when I taught the old emporium model. I knew it was a case of students completing their work independently with the help of the instructor and the tutors and then taking a paper and pencil test. Of course, someone wrote into the grant that the class would be all lab work and no lecture. I guess they did that because they were modeling what another school was doing. That ticked me off.

Professor Milacki's challenge was getting the other instructors on board:

> As chair, I couldn't get anyone to teach these classes. I taught a couple of sections, and after that it was mostly adjunct and limited contract faculty. Most of the full-time faculty thought the layout of the emporium model was beneath them. One faculty member said to me, "I didn't get a master's in math to become a [expletive] tutor."

Professor Mussina concurred:

> I didn't want to teach math this way [the emporium model], but I was the new guy, so if I wanted to get tenure, I needed to do what I was told.

Professor Lopez added:

> The emporium model worked for some students, but I didn't like teaching that way. A lot of the fun of teaching is creating lessons and activities to help students learn. All I did in the lab was walk around and answer questions. I felt more like a glorified tutor than a college professor.

The emporium model was met with some positive results from the BCC students. Professor Lopez shared:

> Some students honestly liked it. We found the students who were advised correctly from their counselors and knew that they were signing up for a course in a computer lab were happy with the modality. They also liked that they could work at their own pace. We gave them pre-tests before

each module, so they could test out of certain modules if they knew the material. We also had the computer lab open all day, so that they could come in at any time to get help.

Professor Lopez also discussed some negative results:

> We had some students who hated it. They were surprised and over-whelmed that they had to learn math in a computer lab. This was especially the case for some of the older students, who were not as computer literate. It took forever for some of them to register with MyMath Lab on the first day. Some of them just got up and left the first day in frustration.

Kilgus Community College also employed an emporium model during the aughts. However, this modality blended lecture with self-paced practice.

> I would lecture for 20 minutes and then let the students work on problems. I felt it was necessary to do some teaching for the students. I never wanted to do the straight emporium model with no lecture, explained Professor Johnson.

GCC also started using an emporium model that blended lecture with drill and practice for some of the developmental math classes in 2006. This was a faculty-driven initiative led by a senior GCC faculty member. Professor Sutcliffe explained:

> It was basically a hybrid model. The class took place in a computer lab using MyMath Lab, and the students could work at their own pace, but there was also a lecture component.

Unfortunately, this hybrid model led to complications:

> We had about 10 or so sections of those emporium model classes, and each class ran for 70 minutes three times a week. The problem was some instructors lectured for 20 minutes; some for 30 minutes. Others for 40 or 50 minutes. Some lectured the entire time, because they just didn't like the lab model. Some didn't lecture at all. It was a mess.

Some mathematical topics were more conducive to student learning in the emporium model:

> I think the arithmetic classes worked well in the emporium model, *reflected Professor Sutcliffe*. Those concepts are simple and don't require a lot of steps. For many students, that is refresher material.
>
> I always dreaded teaching arithmetic in the regular classes, *shared Professor Mussina*. You start talking about place value, fractions, and decimals, and the students look at you like you're crazy, because it's such basic content. They really appreciated being able to work through the content at their own pace without being lectured to.
>
> What I always like about the emporium model was that students got a lot of drill and practice with difficult topics. After all, you get better at math by doing lots of problems. So, topics like fractions and the laws of signed numbers, that are such foundational topics for math, I think it benefited the students to just do so many problems, said Professor Milacki.

However, other mathematical topics were complicated by the emporium model.

> Anything that took a lot of steps and a lot of organization was a tough topic to teach in the lab, *conveyed Professor McDonald*. Equations and especially equations with fractions were hard. Students really need to organize their work, and it's hard to teach organizational skills in the lab. You do so little guided practice, and it's during the guided practice where you help students with organizational skills.
>
> Factoring using trial and error was really tough in the lab, *shared Professor Lopez*. This was especially the case when the leading coefficient is greater than one. Students want to rush to the answer, but they don't want to organize their work. With so many students working at their own pace, it was hard to work with them to organize their work.
>
> What I like about lecture or guided practice is I can introduce the topic and more importantly show them good organizational skills. Since I wasn't introducing the content, students were just seeing the problems on MyMath Lab and working on them. When I would come around to answer questions, it seemed like they already developed bad organizational skills, and I was a step behind, *summarized Professor Milacki*. Basically, in the lab, they were developing bad habits before I could even teach them.

The goal of the emporium model was to accelerate students through their developmental math sequence at a quicker rate. Did the faculty members see an increase in acceleration?

> No, not really, *said Professor DeLeon*. We gave students the opportunity to work ahead. We even designed a schedule so they could complete both basic arithmetic and the algebra 1 class in one semester, but most of them did the least they could. It was hard just to get them to do the basic required work.
>
> We had a few students who worked ahead and finished the basic arithmetic and algebra 1 class in one semester, but they were anomalies, *recalled Professor Sutcliffe*. They were students who were strong in arithmetic and then spent a lot of extra time in the computer lab working on their math.
>
> I think on average, we had about five percent of the students who started in the basic math class and completed both courses in one quarter, *shared Professor Lopez*. These were usually the higher-end students. They could see an arithmetic concept, and it would come back to them really fast. They also worked hard and spent a lot of time on their math.
>
> It was very few students who could complete two courses in one quarter, *shared Professor Mussina*. I'm still glad they had the opportunity. Some students would have been bored in a regular lecture-based class.

The Return of Learning Communities

Learning communities experienced a resurgence in the late aughts, and this was particularly the case in developmental math classes where community colleges were frantically seeking ways to increase student success rates. Like the learning communities model from the 1970s (as discussed in Chapter 2), students registered for two classes as a cohort. Advocates of learning communities asserted that that the initiative would produce two positive outcomes. First, students would develop meaningful connections with both their peers and their instructors. This was based on the data that students are more likely to be successful and persist in college when they make meaningful connections with their peers and their instructors (Tinto, 1998). Second, learning communities would enrich academic instruction. Like the 1970s model, traditional lecture was supplemented with collaborative learning. Additionally, the instructors of each course were encouraged to help students understand common themes between the two courses. Also, like the 1970s model, each college was assigned a learning communities liaison, which was a person who was paid for overseeing the implementation of the initiative (Visher et al. 2010).

The Learning Communities of this era experienced some positive results. Some colleges reported higher overall success rates in developmental mathematics courses that employed learning communities compared to those that did not. Several students conveyed that they felt more support in both their academic and personal lives. More specifically, students felt that the learning communities helped them establish more personal connections and enhanced their academic progress (Visher et al., 2010, Weissman et al. 2011).

There were, however, some negative findings from learning communities. While students in developmental mathematics learning communities had higher success rates in their specific math courses, they were no more likely to persist in their college career than their peers who were in a non-learning communities model (Weissman et al. 2011). Colleges also struggled to increase the number of learning communities due to complications of students' schedules. It was difficult to find enough students to register for two of the same classes. (Ashburn, 2007). It was also challenging to find common themes between two separate courses (Visher et al., 2010).

Habyan Community College (HCC) employed learning communities in the late 1970s and phased them out in the 1980s. However, in 2005, as part of the Achieving the Dream initiative, the institution brought them back. Professor Morgan recalled:

> I was excited when they [learning communities] returned. I had good memories of them, and I liked how it help the students connect with each other.

However, immediately, Professor Morgan could tell it was different this time.

> Last time, it was the faculty who decided to implement learning communities. This time, it was a mandate from our dean. In one of her meetings when she was reminding us about our poor success rates, she asked us, "What do you think about learning communities?" Next thing, I knew, we were doing them. I was eager to help in a leadership role, but she said we had a college-wide learning communities coordinator who would be overseeing them. That was the case before, so I didn't think too much of it.

Professor Morgan noted another difference:

> When we did learning communities in the [19]70s, I liked the collaborative learning aspect, but it was different this time. Before [1970s], we would teach the material to help them [the students] understand it. Then, we would have them to work together and help each other. This time, the learning communities coordinators wanted us to use kind of an inquiry approach.

As discussed in Chapter 4, in inquiry-based instruction, students are expected to work together in groups to master math concepts. The instructor acts as a facilitator, who asks thought-provoking questions. Learning takes place through discussion and collaboration as opposed to lecture. Unfortunately, learning communities soon became a nightmare.

> It was a mess. First of all, it wasn't like before where the [learning communities] coordinator simply helped us. This time, we were being micromanaged. The coordinator wanted to see our syllabi and told us what we should and should not have in there. We also had to have a common theme in the courses we were paired with. In one case, my introduction to algebra class was paired with a cooking class. Yes, you use math in cooking, but it was still hard to find common themes that students could understand.

This created a great deal of confusion for the students.

> We spent so much time trying to help students find themes that they couldn't master the required material. The students also hated the inquiry approach. They struggled to understand the material without my direct guidance. Between being micromanaged by the coordinator and dealing with confused students, it was not a good experience.

Professor Morgan summarized:

> It was so much simpler when we did this [learning communities] in the 1970s. We just helped students make connections with each other. This time, the administration complicated things for no reason.

In 2006, TCC implemented learning communities as part of the Achieving the Dream grant as well. Like the faculty at HCC, the faculty at TCC were strongarmed into the initiative. Professor Mesa explained:

> Our dean said we were doing them [learning communities] so that was that.

Professor Mesa noted immediate problems.

> Our [learning communities] classes were out of control. I always had a lot of behavioral problems in those classes. I think the students were frustrated by trying to find the common themes and all the required group work. My basic [concepts of] algebra class was paired with a sociology class. There were no common themes.

Professor Bell added:

> The students were frustrated because they struggled in math, and they wanted us to explain the material to them. But our administration mandated that learning communities be taught using group-based instruction, and for students who already had math anxiety, it [group-based instruction] created more anxiety and frustration.

Learning communities at TCC were short-lived.

> Our dean wanted us to always have like 20 or 25 learning communities pairings. I guess that was part of the stipulations of the grant. Students have such complicated schedules that it was hard to get them to register for both classes. We were lucky if we got 10 learning communities to go.

Professor Mesa recapped:

> I needed to get students to understand equations, inequalities, and polynomials, but I was required to have them work in groups, teach each other, and try to make connections with all these themes. The bottom line is the learning communities took away from quality instruction.

Both HCC and TCC witnessed abysmal success rates with learning communities.

> My success rates with the learning communities sections were always worse. The students just weren't learning anything, recalled Professor Mesa.
>
> I had one section where less than 10% of the students passed, shared Professor Morgan.

Lecture Becomes a Four-Letter Word

In the 1990s, battles broke out in the K-12 field over effective pedagogy. While some educators believed that lecture or guided practice was a prerequisite for applying the concepts, progressives asserted that students learned better through collaborative learning practices such as inquiry and group-based instruction as well as project-based instruction. In the 2000s, these battles made their way into community college math.

The aughts were clearly a time of attempted reform in community college math. In addition to pushing acceleration and improved success rates, college administrators also began to discourage lecture-based instruction and push more inquiry and collaborative leaning in mathematics, especially introductory mathematics courses. This was the case in the learning communities initiative, as mentioned in the above section. The driving force for this push was not clear. Certainly, there are statistics that show the benefits of collaborative learning. Some faculty offered various thoughts as to this change in mindset:

> Our Achieving the Dream leader basically stood up in front of us [the developmental math department] and told us that traditional lecture wasn't working for us, so we needed to try something else, shared Professor Mesa.
>
> It was monkey see, monkey do, *recalled Professor Ballard.* Someone checked out another school [community college] and they were doing inquiry for their developmental math classes, so we had to do it.
>
> These discussions were happening at conferences like AMATYC, *said Professor Thurmond.* Collaborative learning, inquiry-based instruction, and learning communities were viewed as innovative, and our success rates were so bad, we had a hard time defending the traditional lecture-based instruction.

Colleges were required to employ inquiry-based instruction as part of their change initiatives. Professor Ballard elaborated:

> We were part of the Achieving the Dream initiative. That meant we had to look at our abysmal success rates in our developmental math [classes] and introductory college-level math classes and figure out how to change them. I still don't know who, but somebody decided we were going to try inquiry-based instruction for our algebra 1 class. I still think it was Preppy Harvard [Professor Ballard's division dean].

Professor Thurmond discussed how Griffin Community College was mandated to implement inquiry-based instruction into one of their developmental math classes

> We had a faculty member who went to AMATYC, and he went to one of those concurrent sessions on group-based instruction and he drank the Kool-Aid pretty fast. He came back and was trying to convince everyone we had been teaching our students the wrong way and this [group-based instruction] was the way to go. He convinced our dean, and guess what? My dean wanted us to try group-based instruction for developmental math, so I had to get the faculty on board.

Inquiry-based instruction was a short-lived disaster.

> We tried the inquiry instruction around 2005 or 2006. It was such a disaster, we dropped it after one academic year. We had this really annoying Achieving the Dream leader who micromanaged us while we tried implementing inquiry into our algebra 1 class, shared Professor McDonald.

Professor Thurmond, who had to employ this method in the basic arithmetic 2 class, elaborated on the problems that occurred:

> Let's see. The students complained from day 1 that they had to work in groups. Most of the time, I had students who were frustrated that they just weren't understanding the content. Yes, it was pretty easy content, but the reason most of our students test into this class is that they don't understand concepts like fractions, and decimals and proportions. Students were complaining in every one of those sections [inquiry-based instruction] that they could not learn this way. Students also argued with each other and complained that people in their groups weren't doing the work. I was more of a mediator than a teacher. Our success rates in the basic arithmetic 2 class dropped even further. After a while, I just told my faculty to stop doing this.

Like the emporium model, the faculty were concerned with how inquiry-based instruction negatively impacted their students' organizational skills.

> Students placed into the basic arithmetic [2] class because they lacked organizational skills. When they attempted to try these problems on their own, they were all over the place, said Professor Sutcliffe.
>
> Solving equations are so important for success in math. I mean, they need equations for as long as they take math, and it takes a certain skillset. Equations with fractions are especially hard, and students need to be very thorough when organizing every step, *said Professor Ballard*. Letting them try and do those without my direction set us even further behind, because I would have to correct even more bad habits they developed.

When Professor Sutcliffe taught elementary school, she studied and utilized inquiry-based instruction. She stated some prerequisites for inquiry-based instruction to be effective:

- There needs should be teacher buy-in. More specifically, the teacher must believe this method will help students learn.
- Students should have some experience with inquiry-based instruction. Moreover, the students need to buy into this type of learning.
- Before utilizing inquiry-based instruction, teachers need to assess their students' skill level.
- Students should have some foundational knowledge regarding the topic.

Professor Sutcliffe opined why inquiry was poorly implemented and a poor fit for developmental math students:

> First of all, most of the students never had to learn math through group-based instruction, or it had been a long time since they had to. So, it was a shock to them. Our students have a lot of math anxiety, and they lack

basic foundational knowledge. We need to engage them in their learning, but they need structured instruction. Finally, this initiative was just
thrown on us. Faculty need extensive training on inquiry, and even then
it should be up to the faculty member if they want to use it [inquiry-
based instruction].

Service-Learning: An Effective but Misplaced Initiative

Since the 1970s, service-learning has become prevalent in education. Service-
Learning consists of higher education students receiving academic credit
while providing service to a place of employment or a community organization. Students then learn through experience, application, and reflection
(Kenny & Gallagher, 2002). The concept of learning through experience can
be traced back to the beliefs of John Dewey and other progressives in the
early twentieth century.

The practice of service-learning received increased support throughout the
twentieth century and into the aughts. Advocates of the practice assert that
students who participate in service-learning develop a deeper understanding for an academic discipline. For example, students who serve in a soup
kitchen cultivate more of an understanding of sociological issues such as
hunger, homelessness, and inequality (Brail, 2013). In general, studies have
shown that students who enroll in courses with a service-learning component receive higher grades in their respective courses and persist in college
(Eyler et al., 2001; Celio et al., 2011 Brail, 2016, Nakata, 2020). With community college administrators grasping at straws during the aughts to increase
success rates in developmental math, it is no surprise that service-learning
became an initiative to try and improve the discipline.

SCC required that all developmental math courses employed a service-
learning component in the fall of 2008. This was part of SCC's participation in the Academic Quality Improvement Program (AQIP). AQIP is part
of an institution's accreditation with the Higher Learning Commission.
Departments choose initiatives that will hopefully increase student success
and completion. Consequently, students completing the basic mathematics
classes at SCC had to complete some type of service-learning requirement.
The specifics were left up to their instructors. Professor Ballard discussed the
implementation of this initiative:

> Preppy Harvard strikes again. Here's what happened. We had this guy
> come who was supposedly an expert on community college math. I think
> he was from Texas or something. He met with us as a department, and he
> was making all kinds of recommendations. One was, "all developmen-
> tal classes should have a service-learning component." That's all Preppy
> Harvard needed to here, because next thing you know, we were doing it.

Professor Ballard elaborated:

> There was a service-learning coordinator, who kept telling us how great service-learning was, but she wasn't from our department, and she had no idea how service-learning should be applied to developmental math.

Professor Ballard explained the application of service-learning to the basic mathematics course:

> Students had to complete at least six hours of service-learning. It just had to be volunteer work somewhere, and they had to relate it to developmental math.

The process was chaotic:

> First of all, it was such a low-level class, that we had students who didn't understand what was required of them. We had students complaining about their work schedule and transportation issues. So, they didn't have time to do their service-learning project or have a way to get there. What frustrated me was I spent so much time explaining the project, it was taking time away from me teaching actual math concepts.

What service-learning projects did the students choose?

> You mean the ones [students] who actually completed the project? Because many didn't. Some volunteered at a soup kitchen; some volunteered at a local community that updated housing; some volunteered at the library. Some students even noted how you used basic math like fractions and decimals in construction or measuring food and drinks, but it was minimal.

Professor Ballard discussed the outcome of the service-learning project.

> A lot of faculty dropped it the first quarter. It was too chaotic; it caused too much confusion; students didn't see the purpose, and most importantly, it took time away from our teaching. I don't know or care how Preppy Harvard made that disappear from the AQIP requirement.

Something about the service-learning initiative confused and still confuses Professor Ballard.

> The thing is we were already doing service-learning in one of our higher level math classes. We didn't call it that, but it was service-learning. In our math for future teacher's class [teacher preparatory], we required students to volunteer at a school and basically assist the cooperating teachers. They then had to reflect on their experiences. The hope is this would help them as future teachers. We tried to explain this to Preppy Harvard, that we were already doing service-learning in a math class, and it was working, but it was like talking to the wall; he was insistent that we had to do it [service-learning] in a developmental math class, probably because that guy from Texas said we should.

Professor Ballard offered some final thoughts regarding the service-learning initiative.

> Service-Learning gives students a good chance to contextualize, and in hindsight contextualization was sorely missing from the lower-level math classes in that time period, but it was just a complete misfit for those kinds of students.

Supplemental Instruction

Supplemental Instruction (SI) was yet another practice employed in the aughts to improve success rates in developmental math. While there are variations on the practice, SI starts with hiring or appointing an SI leader for a specific class. An SI leader is generally a tutor who is familiar with the content of the class. An SI leader may even be a student who completed that class. The SI leader attends the class with the students, and then holds additional sessions outside of class where students can obtain extra help with the subject matter (Finney & Stoel, 2010). SI leaders may also receive training regarding working with developmental math students (Wright et al., 2002). Again, the logistics of the SI sessions may vary; however, they are typically held twice a week for 50 minutes (Maxwell, 1997; Phelps & Evans, 2006).

Like learning communities, the employment of SI is a way to build community and help students make connections. The SI leader may employ group discussions regarding a topic. In addition to helping students with the mathematical content, SI can help students combat isolation and help them to realize they are not alone in their struggles (Tinto, 1993; Phelps & Evans, 2006).

The practice of SI was created by Deanna Martin in 1973 at the University of Missouri in Kansas City (Phelps & Evans, 2006). However, SI gained popularity in the aughts, as it experienced success in institutions such as Valencia Community College in Florida. Like the emporium model, several institutions adopted SI as part of the Achieving the Dream initiative. Overall, in the aughts, SI was positively received among students and faculty. Students reported more confidence and lower anxiety in developmental math. These students also showed higher success rates in their classes. There were also debates as to whether SI sessions should be mandatory. More specifically, should students be required to attend SI sessions? Consequently, students who enrolled in mandatory SI sessions had higher success in their developmental math classes than those who did not attend (Phelps & Evans, 2006, Finney & Stoel, 2010).

In 2007, TCC employed the SI model for their basic concepts of algebra class as well as their computational skills course. During the 2007–2008 academic year, TCC tried a non-mandatory version of SI. Professor Mitchell explained:

We wanted to be flexible for the students. So, we hired some SI leaders to run 50-minute sessions. We ran them most of the day, Monday through Friday from like 10am till like 6 or 7 pm. I think we ran over 20 of them each week. Students could drop in for any session they wanted. They didn't need to register or anything. They would just show up with questions on any given topic. The SI leaders had class schedules, so they knew what the current topics were, but they were prepared to answer any questions.

Professor Mitchell shared the positives:

We found that over the first two quarters of optional SI, the students who participated had an 83% success rate in their courses. The written feedback we got from students was great as well. They said the extra help and individual attention they got from the SI leaders really helped them succeed.

Professor Morgan explained the negatives:

Our success rates were good, but I believe we had less than 10% of the students in those courses actually attend those [SI] sessions. We basically learned what we already knew, and that was that developmental students do not do optional.

For the 2008–2009 academic year, the faculty at HCC and TCC decided to implement mandatory SI for developmental math classes. However, this required the faculty to redesign the program. Professor Mesa explained:

When students registered for a computational skills and basic concepts class, they had to register for a one 50-minute SI session, and the session was designated for that class.

This also required assigning specific SI leaders for each class. Professor Morgan explained:

It was a logistical nightmare. We had to make sure there was a session for each class, and we had to do that through registration. We had to make sure there was a specific SI leader for each class and that SI leader could attend the regular classes [computational skills and basic concepts of algebra]. We also worked to help the SI leaders create class plans.

Overall, the SI initiative was received well by the students.

The students really liked the [SI] sessions. They appreciated the extra time to ask questions and get more help with topics. This was especially the case when we covered hard topics like equations, inequalities, and polynomials, shared Professor Mesa.

The students seem to really make connections with each other, more than they did in the regular classes. They [the SI leaders] did a great job of engaging the students. They would really get the students to open up about their struggles. They would also have the students work together on problems, recalled Professor Mitchell.

TCC witnessed a slight increase in student success by using SI but nothing significant.

> The students who took advantage of SI did better in their courses, but we still had attendance issues. After a while, many students would blow off the SI sessions, and believe me, there were students who needed to attend those sessions who didn't, *recalled Professor Morgan*. We just didn't have a way to require them to go. You would think they would just show some student responsibility, right?

The faculty also noted another issue with SI. Professor Mesa explained.

> The computational skills course was a big problem. We had students in there who didn't know how to divide; they didn't know how to multiply; some didn't even have basic number sense. Our administration thought that SI would help these students, but you know something, 50 extra minutes a week wasn't going to help these people.

Professor Mitchell added:

> When you have students who are so far behind and have so many gaps, giving them one fifty-minute session doesn't cure everything. I had students in my basic algebra class who were still struggling with the laws of signed numbers. When you don't understand the rules of signed numbers, you can't evaluate expressions, you can't solve equations, you definitely can't solve inequalities. One SI session wasn't going to catch them up.

Distance Learning Morphs into Online Learning and Explodes

Distance learning in mathematics grew in the 1990s; however, the practice soared in the 2000s. Nationally, from 1999 through 2005, the number of students taking at least one online class rocketed from 744,000 to over 3 million (Weisbrod et al., 2008). This was the case for math classes at HCC and GCC.

> I didn't want anything to do with distance learning when it first came out, *shared Professor Bell*. It just seemed so hard and impersonal. I got into it around 2006. I really liked MyMath Lab. I liked how interactive the software was, and how user friendly it was, and we started doing most of our online math classes with it.

Professor Sutcliffe also liked the interactivity of MyMath Lab.

> I was hesitant about teaching online, but I liked how students could get immediate feedback when they answered a question, and there were all those resources like the videos and how students could email their questions to me through MyMath Lab.

The faculty at SCC sought to serve their distance learning students; however, they realized that their methods were outdated. Professor Trombley elaborated.

It was the 2000s, and we were still using the VHS tapes. Students were still mailing in or dropping off written homework. We were looking for ways to update our teaching methods. So, we started using MyMath Lab, so we could see students' responses, and they could share their questions with us.

However, the SCC faculty felt that MyMath Lab lacked the thorough instruction that students needed.

MyMath Lab had these very short videos, but the students needed more, especially the developmental math students and the algebra 2 and algebra 3 students. So, we started creating instructional videos using Vimeo.

It was another example of faculty helping each other that fortified this initiative.

I knew how to create videos, *said Professor Trombley*. I'm a geek. I helped some other faculty in the department create some videos as well.

The faculty at HCC felt the same way regarding thorough instruction for distance learning students.

Students need detailed instruction. We were using ALEKS software, which was interactive, but we wanted to be able to explain problems to students and also help them to develop organizational habits, *said Professor Morgan*. It was actually around the time You Tube first came out. One of our young teachers knew how to make videos and post them on You Tube. Our students really appreciated it.

The Challenging Student Population

Since the inception of the community college, faculty have faced challenges with students. Students have demonstrated affective behaviors such as poor attendance and subpar work habits and haphazard study skills. Weak arithmetic skills have also been an issue. Students in developmental classes have struggled with concepts such as fractions and decimals and even overall number sense. However, the faculty noted that the student population became more challenging in the aughts. It seems cliché that teachers note that students become more challenging over time. As Professor Johnson noted:

When I first started teaching, the older teachers said that the student population had changed and had gotten worse since when they first started. It seems like all teachers say that the students were better when they started.

Nevertheless, several faculty participants noted specific issues with students that became more challenging in the aughts.

Higher Demand for Arithmetic Courses

Again, struggles in arithmetic have been problematic since the 1970s. Nevertheless, this became more challenging in the aughts. Professor Ballard elaborated:

> We always had students who didn't get fractions, decimals, percents, and even multiplication and division, but it seemed like after 2000, there was just more of them. The number of sections of basic mathematics classes we offered in 2005 was up by 70% from 1995. By 2009, it was up by another 50%. So, it was still the same old arithmetic issues. We just had more students struggling with them.

Professor Morgan added:

> I think by the 2000s, we had so many more students coming back to school. When I started in the [19]70s, it was primarily White students, 18 or 19, right out of high school. By the 2000s, we had White, Black, Latino, and all different ages. So, I just think more students in an open admissions school led to more of a need for basic math. That's especially the case for students who had been out of school for many years and then take a math placement test.

However, Professor Mitchell felt the increased need for basic skills had more to do with the varied avenues that brought students to community college:

> When I began teaching, they [the students] came from high school, and yes, many had deficient math skills because they didn't take much math in high school, but by then [the 2000s], we had students on work release programs from prison; we had students who had been released from mental institutions; we had students who simply didn't know what to do, and someone told them to sign up for community college because they get financial aid. So, you can see how learning math hadn't been their top priority.

Professor Sutcliffe concurred:

> Around 2006, I was on a date with this woman who was a counselor at a juvenile detention center. I told her that I taught at Griffin [Community College] and she said, "You know, you get a lot of our students. When they leave us, we recommend them to you all."

The abysmal skills frustrated and stunned some faculty:

> I just couldn't wrap my head around students not understanding how to add or subtract, *recalled Professor Mussina.* But that is who I got in the basic arithmetic 1 class. How did these people get through high school and life?
>
> We had so many students who didn't know their math facts, like addition and multiplication facts, *stated Professor Thurmond.* I don't think rote memorization should be the sole source of instruction in

math, but you need to understand basic facts to be able to do any kind of math. When someone asks you what is nine times three, you shouldn't struggle.

The growing need for arithmetic instruction created staffing issues.

By the late 2000s, the country was dealing with the Recession of 2008. Our enrollment overall was going through the roof, *recalled Professor Bell.* From 2008 to 2009, our enrollment at Habyan [Community College] went up by 25%. We were getting a lot of students from these displaced workers programs, and many of them were testing into basic mathematics. In fact, by 2009, that class had the highest enrollment in the college, shared Professor Bell.

So, how did this create a staffing issue?

As a chair, I couldn't find enough people to teach that course. Most of the full-time faculty didn't like to teach math that low, and there were only so many adjuncts.

Professor Thurmond, a chair at this time, concurred:

I had a lot of sleepless nights before the semester started, and that had a lot to do with that basic arithmetic class. Even a few days before school started, I would be in my office calling people to see if they can teach a section of that course. I was literally scraping the bottom of the barrel just to get warm bodies to teach a course with content so many students struggled with.

Equity Concerns

Throughout the aughts, community college math faculty continued to get bombarded with demoralizing statistics regarding student success rates. Such statistics extended to equity issues as well. Academic researchers began reporting that there was an overrepresentation of Black and Hispanic students in developmental math classes (Hagedorn, et al., 1999). More specifically, 46% of Black students and 51% of Hispanic students enrolled in at least one developmental math class. Conversely, only 31% of White students and 29% of Asian students tested into a developmental math class (Adelman, 2004). Additionally, White students were 60% more likely to pass a developmental math class than Black or Hispanic students (Bahr, 2010).

The faculty in this study testified that equity concerns were brought to their attention.

By around 2006 or 2007, it felt like as math faculty we went from one crisis meeting to another. It was always the same thing. Too many students were in our developmental math classes and too many were failing, *recalled Professor Mesa.* And on those dreadful reports, we would see that our African American and Hispanic students were really struggling.

> I remember during one meeting, this project leader, you know some administrator who was put in charge of helping us get our success rates up, told us how low our developmental and college level math success rates were for Black students, *shared Professor Trombley*. She asked, "have you thought about what you can do about that?"
>
> This [equity concerns] came up a lot at our school, *reflected Professor Sutcliffe*. It was always being brought to our attention, but no one had any idea how to help.

Prior to 2000, there was not much data collected on community college math students; however, I asked the faculty if they noticed this equity disparity before 2000.

> To be honest, it's [the equity gap] not something I ever noticed till it was brought to my attention. I can say that in the 1970s, my classes were generally White students who were 18 or 19 years old. As time went on, the student population became more diverse, reflected Professor Milacki.

Professor Morgan, however, did recall when she noticed a sharp change in the developmental math student population.

> It was the late 1990s. We got so many students from the inner-city schools. These students graduated with very low skills, and someone said, "Hey send them to community college." So, we got a lot more Black and Hispanic students and many of them had very weak academic backgrounds.

Additional Behavioral and Social Issues

Even back in the 1970s, faculty reported various student behavioral issues such as talking in class or simply not paying attention to instruction. However, by the 2000s, these classroom issues had escalated:

> We were getting students, especially in the developmental courses, who had severe emotional issues, recalled Professor Milacki.
>
> It was in 2004, I believe, I actually had a student go off on me and curse me out in front of me class, because she couldn't get how to multiply fractions, shared Professor Thurmond.
>
> I started getting people drunk coming into my classes. I never thought I would see that happen, stated Professor McDonald.
>
> I would recommend students to see the counselors because of all kinds of behavioral problems, *said Professor Mesa*. When I would follow up with the counselors, they would recite a laundry list of emotional problems they had.

Professor Ballard offered his thoughts for some of the severe learning and behavioral issues that increased during the aughts:

I saw this coming way back in the [19]80s when we had the crack epidemic, and there were all those crack babies. I knew when they turned 18, they had to come somewhere, and we are open admissions institution that offers remedial education. Connect the dots.

Shorter Attention Spans

Several faculty noted that as the years progressed, students' attention spans waned.

Getting students to pay attention in math was never easy, but by the 2000s it was different. They had trouble focusing in class, but so many students lacked the discipline to even complete one algebra problem, reflected Professor Morgan.

Completing a math problem takes a lot of concentration and discipline, and more and more I could tell students would get easily distracted, stated Professor Thurmond.

It became more difficult for students to sit through a two-hour class and maintain their focus, *shared Professor Bell*. It was a signal to me that I needed to mix up my classes more, you know group work, allowing them to work on problems themselves, but still, they just lacked the discipline to do math.

Professor Mesa provided an offering:

The times were changing. The students we were getting grew up playing video games where they did all this multi-tasking. They were watching all those reality TV shows with no linear plots and the camera would just bounce around. Heck, students were texting on their phones while carrying on a conversation. I'm aging myself, but I grew up with board games, very simple video games and watching half-hour episodes of Threes Company.

Professor Mesa noted how this change negatively impacted learning math.

Look, as teachers we need to adjust to the times and ultimately find ways to engage our students, but math requires concentration and focus. I worry that as multi-tasking becomes more of a societal norm, it will make learning a discipline like math more difficult.

Student Entitlement

Faculty noted that during the aughts, students displayed an increased sense of entitlement.

Students thought because they paid for the class, or somebody did, that they should pass, *shared Professor Trombley*. I would have students come to class, not do the work and fail the tests and be genuinely shocked they failed.

It seemed like when students tested into developmental math, they were more resentful that they had to be there. On some level that was always the case, but I noticed it more in the 2000s. It was like they recognized the content, things like signed numbers, evaluating expressions and equations. Even though they didn't know how to do them, they remembered being taught this, and consequently, the college was doing them a disservice by making them take a developmental math class, reflected Professor Timlin.

Some faculty asserted that student entitlement stemmed from bad habits from secondary education.

Here is the problem. In high school, students can be failing their classes, but they always seem to pass. So, when they come to us [community college], they think they will somehow pass, shared Professor Lopez.

I can't tell you how many students I've had who fail exams, miss classes, not do homework, and they think I will give them some magical extra credit project that will give them a passing grade, stated Professor McDonald.

Professor Sutcliffe spoke from experience as a former high school teacher:

Failing a student, especially a senior [in high school] is not easy. Generally, when a student is in a position to fail a class, you'll get parents, administrators, even board of education members who pressure you to pass the student. There were several times I was on the verge of failing a student, and my principal instructed me to give the student extra work to pass my class. It's no wonder our students are shocked when they fail.

Professor Lopez summarized why entitlement issues can serve as a barrier to learning math.

The bottom line is math, at any level, can be a difficult subject to master. It takes a lot of discipline and organization, but most of all, it takes the right kind of attitude, a positive attitude. Going into a math class with entitlement issues gets you off on the wrong foot right away.

Low Admissions Standards on a National Level

While many community colleges had been open access since the 1960s, national data during the aughts suggested that there were particularly low admissions standards in community colleges. For example, in the state of Arizona alone, between 1999 and 2011, 3,292 high school diplomas were awarded to students who were classified as mentally retarded. There was nothing that restricted these students from entering community college (Scherer & Anson, 2014). Additionally, students were entering community college without a high school diploma on a national level. In 2006, *The New York Times* reported that nearly 400,000 students who were enrolled in community college did not possess a high school diploma or equivalency degree (Arenson, 2006). The community college student population had always been challenging; however, the aforementioned data shed light on why this population became especially challenging in the 2000s.

Conflicting Messages from Within

By the 2000s, community college faculty were receiving pressure from their administrators and being barraged with dismal statistics. However, the faculty also reported pressure from their colleagues in other departments such as engineering, chemistry, and physics. More specifically, the faculty from these departments complained that too few students were attempting courses in applied fields because they were not completing their general math classes. It is noteworthy that students who major in the physical sciences or engineering generally need to complete at least two semesters of calculus.

Professor Lopez elaborated:

> I'll never forget it. We were at a meeting to discuss success rates and the chair of an engineering department looked over at us [the math department] and said, "Over the past 15 years, over 800 students, who started developmental math, declared themselves as engineering majors, but didn't complete their math requirements."

Professor Mussina, who was also at the meeting, gave his thoughts:

> I couldn't believe what I was hearing. This guy is an engineer. He knows the kind of students who test into developmental math. He knows the rigors of calculus. Why is he blaming us?

Professor Bell stated that the math faculty were receiving similar pressure at Habyan.

> This one physics professor came to one of our department meetings. He was basically talking to us about how to get more students through their math classes and into the applied sciences. He told us we were being too formal in our teaching. We were making our students memorize too many algorithms and formulas in algebra. He encouraged us to be more informal and casual in our teaching. He told us how when he needed to help his own students in math, he would simply show them how to do a problem without stressing the organizational skills and linear fashion we did.

Professor Bell was perplexed by this recommendation:

> We [the math department] knew what he [the physics professor] wanted. He wanted us to pass more students, so that the enrollment in the physics department would increase. But I felt like he was a hypocrite. Students who take physics need to take calculus. There is nothing informal or casual about calculus. Understanding limits, derivatives, and integration requires a solid understanding of all algebra plus other math. Oh, and you need good organizational skills for calculus.

Professor Moyer discussed the frustration from conflicting messages.

> On the one hand, our colleagues in the engineering and chemistry departments wanted us to pass more students, so they would have higher enrollment, but at the same time they complained that students didn't possess the necessary math skills. One time, a chemistry professor stopped me in the hall to talk. After a few minutes, it felt like she was cross examining me on how I teach math word problems.

Professor Trombley concurred:

> I get it. I teach calculus, and the majority of the students fail because their math background isn't deep enough. They don't have deep enough of an understanding of algebra, trigonometry or just more abstract concepts like interpreting graphs. That's why I don't get the push to accelerate underprepared students into higher level math.

Professor Fenimore summarized:

> It was like, you're damned if you do, and your damned if you don't.

Issues with Uniformity and Too Much Uniformity

During the aughts, administrators began noting the lack of uniformity in developmental math classes and attributing such lack of uniformity to the poor success rates.

Exams

Faculty noted that they became under pressure to develop standardized final exams, especially for developmental math classes, to ensure that all students were receiving the same exit assessment.

> It was a painful process. It was hard to come up with a standardized exam, because we all had different ideas as to what should be on the exam for the computational skills and basic concepts course, and it was

my job to get everyone to come to a consensus. It was like herding gold-fish, shared Professor Mitchell.

When my dean told us we needed to come up with a standard final for our basic arithmetic and algebra 1 class, I hated the idea, *reflected Professor Milacki.* I asked the faculty to share their current finals, and I realized we were all over the place in terms of the types of questions we were giving, and I realized we needed a standardized final.

Professor Milacki cited some examples:

So, for the algebra 1 finals, some instructors were using a lot of linear equations but hardly any or no factoring or laws of exponents. Other instructors were over-emphasizing laws of exponents and factoring but hardly any linear equations. Some instructors had no word problems whatsoever. I'm all for academic freedom, but the goal of algebra 1 is to prepare students for algebra 2, so by being all over the place, we were doing our students a disservice.

Professor Bell was in a similar position in getting the faculty to collaborate in developing a standardized final for the basic mathematics, introduction to algebra, and elementary algebra classes.

The key was making sure that all faculty felt included in developing the final but also ensuring that we had an even distribution of questions. As a group, we decided how many of each type of question should be on the final. So, for our introduction to algebra, we decided on a 30-question final, and we would have five problems with factoring, five with the laws of exponents, four with evaluating [expressions] and so forth. Then, I assigned commit-tees to work on each final. Within the committees, faculty submitted ques-tions they wanted for each topic and the committees decided. When all the committees were done, we met as a department and reviewed the exams.

Other faculty noted that uniformity with flexibility was imperative.

It was a tough transition to get everyone to agree on a common final, so we decided that 20 questions would be standardized, and the faculty could create any five questions they wanted. I think just the fact that faculty had the flexibility made it easier to come to consensus for the common final, recalled Professor Thurmond.

Collaboration

In 2008, GCC's developmental math and math departments merged into one department. The GCC faculty members discussed how this created better alignment for students.

I was against it [the merger] for many years. I always thought the depart-ments needed to be separate, because we understood how to teach devel-opmental students, but I think this was best in the long run. The reality is there needs to be alignment between developmental math and math, and that can be done best if we are all in the same department, shared Professor Thurmond.

Professor Timlin added:

We're on the same page more regarding testing policies and calculator policies. We have a math curriculum that just flows better because we're all on the same page now.

Professor Sutcliffe shared:

It was just more of us working to help students. Rather than developmental math faculty and math faculty working separately, we all work together to help students and shared ideas with each other.

Calculators

In the 1990s, there was inconsistency regarding calculator usage. However, throughout the aughts, the faculty noted the importance of uniformity on this important issue.

Maybe we just got tired of students and others upset that we didn't use the calculator. Maybe we just started seeing the benefits of the basic calculator, but by 2006 or so, we allowed the basic [four function] calculator for our developmental math courses, stated Professor Milacki.

We had a lot of calculator debates for many years in the developmental classes, but I think faculty turnover helped. Newer faculty were in favor of the calculators, so it just happened over time, stated Professor Bell.

Having a calculator as optional for some instructors didn't work, *shared Professor Morgan*. Some of us really didn't want the graphing calculators, as they were too complex and too expensive, so we agreed on the scientific calculators. It was a happy medium.

Grading

An additional concern from administrators, especially in developmental math courses, was inconsistency regarding grading policies. More specifically, final grading policies needed more stability. Professor Moyer explained:

This was a big debate when I first started. Some of us had 70 or 80% of the final grade allotted for proctored exams and quizzes, but others only [allotted] 50 or 60%, and allotted other portions for homework and class participation and whatnot.

> I really hated this [the push for grading uniformity], *stated Professor Mussina*. Part of the reason I went into community college teaching was because I thought I would get more academic freedom, but this was for the best. When you have some teachers only using 50% of the final grades for exams for an algebra 1 class, and the student gets a "C" in the class, chances are they will bomb the next class. So, we needed to get together on that.
>
> We reached a consensus on the idea that 70 to 80% of the overall grade in developmental math classes or even classes like college algebra would be devoted to proctored exams. That combined with a uniform final gave our students a better chance to succeed in the next course, and we saw slightly better results in our classes, reflected Professor Bell.

Grading uniformity was especially difficult to grasp for the experienced faculty.

> My dean started getting on us about more common [final] grading for the lower-level math courses, conveyed *Professor Milacki*. At first me and the senior level faculty hated the idea. I mean, we were used to the freedom of teaching and grading how we wanted, and I do believe that as a teacher you need freedom and flexibility, but you also need some uniformity for the benefit of your students.

Professor Milacki shared some specifics:

> We had a retreat in the early 2000s, and we discovered in our basic math and introduction to algebra class, we were all over the place. Some were requiring the calculator; some were not. We were all over the place with what percentage of the final grade we made exams. Some of us graded homework; some didn't. A couple of faculty members were even giving open book exams! This created problems for the students as they moved from one math class to another. So yes, while I will fight to the grave for academic freedom, there was a lot of disarray in our department, and it was hindering our students.

Too Much Uniformity

While the faculty agreed that there should be certain elements of uniformity, particularly in developmental math classes, some felt that this push for uniformity went overboard. Professor Ballard explained:

> Preppy Harvard [Professor Ballard's dean] kept trying to get us to teach the exact same way. He wanted us to get together to decide exactly how we should teach topics like factoring or simplifying rational expressions. First of all, it wasn't his place to do that. Second, he doesn't understand that you can't mandate a teaching style.

I felt like that was the plan with making us teach using that inquiry, *shared Professor Mesa.* It was a way to get us all teaching the same way.

It was really sad, *exclaimed Professor Mitchell.* There was a time when we would proudly discuss all the different things we were doing in our class and the ways we were helping our students. By the 2000s, it felt like if we all weren't doing the same thing, we were doing something wrong.

We did so much to help our students, and we tried so many different things, but then it seemed like our dean told us everything we had been doing for 20 years was wrong, shared Professor Fenimore.

To try and prepare students for their subsequent math classes, some institutions employed mastery learning for developmental math classes. More specifically, students had to achieve a minimum grade on their standardized final exam to pass the course. At Telford Community College, students needed to obtain a 70% on the final for the computational skills and the basic concepts of algebra course:

We went too far, *recalled Professor Moyer.* Students would get a 67% on the final and we had to fail them for one question. It was ridiculous.

We would get students with As and Bs going into the final, and they would fail the exam by a couple of questions, *recalled Professor Mesa.* I think it was more test anxiety, you know the pressure of having everything riding on one exam. We dropped this procedure by 2010.

GCC employed a similar strategy for their basic arithmetic and the algebra 1 class. The basic arithmetic class required a minimum grade of 70%, and the algebra 1 class required a minimum grade of 60%.

We just got tired of hearing about how students were passing basic arithmetic and algebra 1 but failing algebra 1 and algebra 2, so we implemented the mandatory passage rate to make sure they doggone knew the math, reflected Professor Timlin.

It [the mandatory passage rate] really didn't make a difference. The success rates in the next math classes didn't budge. We scrapped the idea after a couple of years, shared Professor Sutcliffe.

What about Statistics and College Algebra?

Much of the reform initiatives in the 2000s centered around developmental math classes as well as beginning and intermediate algebra classes. What was going on with college algebra and statistics?

Statistics Moves into the Twenty-First Century

Traditionally, introduction to statistics classes had been lecture-based, and students were required to compute long formulas by hand. In the aughts, faculty attempted to modernize introduction to statistics and bring in more hands-on applications.

> I actually taught statistics my first year at Bordi, and I just thought there was so much more we could be doing with this course. Everything was just lecture, and students were struggling, shared Professor Mussina.

Professor Mussina approached his department chairperson with some ideas:

> First off, I thought we could be doing more hands-on activities, like labs, to help students understand concepts like probability and confidence intervals. I worked with some of the other faculty, and we started using dice, chocolates, and other manipulatives to help explain basic probability, the difference between mutually exclusive events and non-mutually exclusive events and the difference between independent and conditional probability. Basically, we developed activities that faculty could use. These activities really seemed to help out students.

The faculty at GCC began using more hands-on activities in their introduction to statistics classes as well.

> Probability is such a hard and abstract topic for students to understand, *asserted Professor Timlin.* Students don't really get an understanding for what mutually exclusive events and non-mutually exclusive events are unless they see it. The same can be said for with replacement and without replacement.

In Chapter 4, Professor Timlin discussed the struggles acclimating to the graphing calculator. However, he realized their endeavor into technology was rushed.

> I think we jumped into the graphing calculator too fast. As faculty, we needed time to learn it and how we could use it to better teach our students. As a department, we decided to fully adopt the graphing calculator, and we worked together.

The faculty at Habyan and Telford Community Colleges also sought to modernize their introduction to statistics class. Professor Bell elaborated:

> I had been teaching statistics the same way for 20 years, and I figured it was time to make some changes. After some research, my department began using SPSS [Statistical Package for the Social Sciences] software. You can use it to conduct t-tests, binomial tests, and chi-square tests.

Professor Mitchell added:

> SPSS helps the students understand the content more, but it also gives
> them another tool. They use SPSS in future college-level classes, espe-
> cially psychology and sociology classes, so it helps them relate the con-
> tent to real-life applications and prepare them for future classes.

College Algebra Remains Stagnant

Was there much change in college algebra during the aughts?

> Not much, *shared Professor Mussina*. Our college algebra classes didn't
> change at all in the 2000s.
>   The content in our college algebra was basically the same in 2009 as it
> was in the 1980s, reflected Professor Bell.
>   Our success rates in college algebra weren't good. Our success rates
> in statistics were better, but it was more about the introductory courses,
> you know the gatekeeper courses. Administrators were basically con-
> vinced that if students got through their developmental math classes,
> they would stay in school, asserted Professor Mitchell.

However, despite the low success rates in college algebra, the faculty participants noticed the difference in the student characteristics between developmental math and college algebra.

> Students failed college algebra because it has hard content and students
> came in with weak skills, *shared Professor Bell*. But the students just had a
> higher level of maturity.

Professor Moyer clarified:

> Most of the students who struggled in my college algebra class genuinely
> understood that they were underprepared. Some admitted they didn't
> study enough. Others knew topics they would struggle with in advance.
> My developmental math students are certainly underprepared, but most
> oftentimes, it's their lack of maturity of what they need to do to be suc-
> cessful in a math class that gets them [in trouble].

Summary

The aughts were a tumultuous decade for community college math, especially developmental math. Community colleges were under immense pressure to increase student success rates. Consequently, these schools became
slammed with reform initiatives. Some initiatives were well-received and

yielded positive results such as supplemental instruction. Others were misfits such as learning communities and service-learning. Other initiatives such as the emporium model conveyed mixed results.

There was a major push for uniformity as well. While faculty had tried vigorously to help their students over the decades, a lack of consistency existed, particularly in lower-level math courses. Policies with calculators and grading were especially uneven, and this hindered student success. Therefore, faculty worked to gain more consistency. However, the faculty also found that too much uniformity and rigidity hindered student learning.

Overall, the reform efforts in developmental math created poor morale and occasionally led to chaos. While change was necessary, the faculty felt micromanaged and treated unprofessionally. In some cases, administrators seemed to be grasping at any initiative they felt would increase student success. However, in many cases, this hampered student learning. As the aughts concluded, administrators were still under pressure to increase student success rates in math. Faculty felt frustrated, underappreciated, and overwhelmed. Moreover, a record number of students were struggling in math, and despite a surge of initiatives, no one had any idea how to help them.

References

Adelman, C. (2004). *Principal Indicators of Student Academic Histories in Postsecondary Education, 1972–2000*. Washington, DC: Institute of Education Sciences.

American Council on Education (2004). Putting college costs in context [Data file]. http://www.acenet.edu/bookstore/pdf/2004_college_costs.pdf

Apling, R. N. (1993). Proprietary schools and their students. *Journal of Higher Education*, 64, 379–416.

Arendale, D. (2003, October). Developmental education: Recognizing the past, preparing for the future. *Paper presented at the Minnesota Association for Developmental Education 10th Annual Conference, Grand Rapids, MN*

Arenson, K. W. (2006, May 30). Can't complete high school? Go right to college. The New York Times. http://www.nytimes.com/2006/05/30/education/30dropouts.html?pagewanted=all

Ashburn, E. (2007). An $88-Million experiment to improve community colleges. *The Chronicle of Higher Education*, 53(33).

Bahr, P. R. (2008). Does mathematics remediation work? A comparative analysis of academic attainment among community college students. *Research in Higher Education*, 49(5), 420–450. doi: 10.1007/s11162-008-9089-4.

Bahr, P. R. (2010). Preparing the underprepared: An analysis of racial disparities in postsecondary mathematics remediation. *Journal of Higher Education*, 81(2), 209–237.

Biswas, R. R. (2007, September). Accelerating remedial math education: How institutional innovation and state policy interact (Achieving the Dream Policy Brief). http://www.jff.org/sites/default/files/RemedialMath_3.pdf.

Boylan, H. R. (1997). An evaluation of the Texas academic skills program. http://www.thecb.state.tx.us/reports/PDF/0282.PDF.

Brail, S. (2016). Quantifying the value of service learning: A comparison of grade achievement between service-learning and non-service-learning students. *International Journal of Teaching and Learning in Higher Education*, 28(2), 148–157. https://files.eric.ed.gov/fulltext/EJ1111129.pdf.

Brail, S. (2013). Experiencing the city: Urban studies students and service learning. *Journal of Geography in Higher Education*, 37(2), 241–256. doi: 10.1080/03098265.2012.763115.

Celio, C. I., Durlak, J., & Dymnicki, A. (2011). A Meta-Analysis of the Impact of Service- Learning on Students. *Journal of Experiential Education*, 34(2), 164–181. doi: 10.1177/105382591103400205.

Curtis, J. W. (2002). *Student outcomes in developmental mathematics 1994–1995 through 1999–2000. Locust Grove, VA: Germanna Community College. East Lansing, MI: National Center for Research on Teacher Learning.* (ERIC Document Reproduction Service No. ED 459900).

Eyler, J., Giles, D. E. Jr., Stenson, T., & Gray, C. (2001). *At a Glance: Summary and Annotated Bibliography of Recent Service-Learning Research in Higher Education* (3rd ed.). San Diego: Learn & Serve America National Service-Learning Clearinghouse.

Finney, J., & Stoel, C. F. (2010). Fostering student success: An interview with Julie Phelps. Change: *The Magazine of Higher Learning* 42(4), 38-43. East Lansing, MI: National Center for Research on Teacher Learning. (ERIC Document Reproduction Service No. EJ893003)

Hagedorn, L. S., Siadat, M. V., Fogel, S. F., Nora, A., & Pascarella, E. T. (1999). Success in college mathematics: Comparisons between remedial and nonremedial first year college students. *Research in Higher Education*, 40(3), 261–284. https://doi.org/10.1023/A:1018794916011.

Hebel, S. (2010). *State Cuts are Pushing Public Colleges into Peril.* The Chronicle of Higher Education.

Hinds, S. (2009). More than rules: College transition math teaching for GED graduates at the City University of New York. http://www.cccs.edu/Docs/Foundation/SUN/Math%20Paper.pdf.

Kenny, M.E., & Gallagher, L.A. (2002). Service-learning: A history of systems. In M.E. Kenny, L.K. Simon, K. Kiley-Brabeck, & R.M. Lerner (Eds.), *Learning to Serve: Prompting Civil Society through Service Learning* (pp. 15–29). Kluwer Academic Publishers.

Maxwell. M. (1997). *Improving Student Learning Skills* (Rev. ed.). H & H Publishing.

Nakata, S. (2020, April 28). Service learning boosts academic performance and retention rates in collaborative study. WSU Insider. https://news.wsu.edu/news/2020/04/28/service-learning-boosts-academic-performance-retention-rates-collaborative-study/.

Noel-Levitz, Inc. (2006). Student success in developmental math: Strategies to overcome barriers to retention. Iowa City, Iowa: Author Office of Institutional Research and Planning (2007, May). Academic Skills advancement at Ivy Tech Community College. http://www.ivytech.edu/institutionalresearch/special/REMEDIATION_Analysis_Final_Phase_I.pdf.

Office of Institutional Research and Planning (2007, May). Academic Skills advancement at Ivy Tech Community College: Phase 1 report. http://www.ivytech.edu/institutionalresearch/special/REMEDIATION_Analysis_ Final_Phase_I.pdf

Phelps, J. M., & Evans, R. (2006). Supplemental instruction in developmental mathematics (2006). The Community College Enterprise, 12(1). National Center for Research on Teacher Learning.

Puyear, D. (1998). Developmental and remedial education in Arizona community colleges: A status report. Phoenix, AZ: Arizona State Board of Directors for Community Colleges. East Lansing, MI: National Center for Research on Teacher Learning. (ERIC Document Reproduction Service No. ED423931).

Rizzo, J. M. (2006). State preferences for higher education spending: A panel data analysis, 229 1977–2001. In R. G. Ehrenberg (Ed.), What's Happening to Public Higher Education? The Shifting Financial Burden (pp. 3–35). Johns Hopkins University Press.

Scherer, J. L., & Anson, M. L. (2014). *Community Colleges and the Access Effect: Why Open Admissions Suppresses Achievement*. New York: Palgrave Macmillan.

Squires, J., Faulkner, J., & Hite, C. (2009). Do the math: Course redesign's impact on learning and scheduling. *Community College Journal of Research and Practice, 33*, 883–886. doi: 10.1080/10668920903149723.

Tinto, V. (1998). Colleges as communities: Taking research on student persistence seriously. *The Review of Higher Education, 21*(2), 167–177.

Twigg, C. A. (2011). The math emporium model: Higher education's silver bullet. *Change: The Magazine of Higher Learning, 43*(3), 25–34.

Visher, M. G., Schneider, E., Wathington, H., & Collado, H. (2010). Scaling up learning Communites: The experience of six community colleges (National Center for Postsecondary Research). East Lansing, MI: National Center for Research on Teacher Learning. (ERIC Document Reproduction Service No. ED509307).

Waycaster, P. W. (2001). Factors impacting success in community college developmental courses and subsequent courses. *Community College Journal of Research and Practice, 25*(5/6), 403–416. doi: 10.1080/106689201750192256.

Weisbrod, B. A., Ballou, J. P., & Asch, E. D. (2008). *Mission and Money: Understanding the University*. Cambridge University Press.

Weissman, E., Butcher, K. F., Schneider, E., Teres, J., Collado, H., Greenberg, D., & Welbeck, R. (2011). Learning communities for students in developmental math: Impact studies at Queensborough and Houston community colleges (National Center for Postsecondary Research). National Center for Research on Teacher Learning.

Wright, G. L., Wright, R. R., & Lamb, C. E. (2002). Developmental mathematics education and supplemental instruction: Pondering the potential. *Journal of Developmental Education, 26*(1), 30–35.

6

The Teens Part 1: Turbulence and Change Continue

The early 2000s were a turbulent decade, especially for developmental math. This discipline was hit hard with reform efforts to increase student success. As the aughts transitioned to the teens (2010–2019), the difficulties showed no signs of lessening. Developmental math courses, as well as college algebra and introductory statistics courses, would be hit with additional changes over this decade as well.

New and Returning Participants

In this chapter, Professors Thurmond (Griffin Community College), Bell (Habyan Community College), McDonald (Griffin Community College), Mesa (Telford Community College), Sutcliffe (Griffin Community College), Moyer (Telford Community College), Trombley (Sisco Community College), Lopez (Bordi Community College), and Mussina (Bordi Community College) shared their experiences of teaching community college math in the teens. Additionally, new faculty members contributed their knowledge as well:

Professor DeSilva

Professor DeSilva began teaching full time at Sisco Community College (SCC) in 2010. She teaches both developmental and college-level math. Professor DeSilva earned a bachelor's and a master's degree in math.

Professor Guzman

Professor Guzman started teaching both developmental and college-level math at Bordi Community College (BCC) in 2013 as a full-time instructor.

DOI: 10.1201/9781003287254-6

She holds a bachelor's degree in math and master's degree in math education (18 credits in pure math).

Professor Holton

Professor Holton's teaching career began at Griffin Community College (GCC) in 2006 as an adjunct instructor. He achieved a full-time position in 2010 and has taught developmental math since. Professor Holton received both a bachelor's degree in elementary education with a math concentration and a master's degree in education.

Professor Hickey

Professor Hickey started teaching developmental and college-level math full time at Habyan Community College in 2011. He earned a bachelor's and a master's degree in math.

Professor Williamson

Professor Williamson started teaching at Habyan Community College in 2009 as a full-time instructor. He has a bachelor's and master's degree in math and has taught both developmental and college-level math courses.

Professor Noles

Professor Noles began her teaching career as an adjunct instructor at Telford Community College (TCC) in 2007 and started teaching full time in 2013. She teaches both developmental and college-level math. She obtained a bachelor's and a master's degree in math.

Professor Douglass

Professor Douglass began teaching developmental and college-level math at SCC in 2009. He has a bachelor's and a master's degree in math.

The Ambiguity of Intermediate Algebra

Since the inception of community colleges, students have needed to complete college algebra or introductory statistics courses for transfer credit to 4-year programs or universities. Although courses such as intermediate algebra or elementary algebra did not transfer, they often satisfied the requirements of various 2-year degrees. Consequently, a student's completion of elementary or intermediate algebra satisfied an associate degree. This changed, in some states, during the teens as various states began to declare that intermediate algebra was not college-level math. This modification happened because the content in these classes is covered in high school and is considered review at the college level. Therefore, some public colleges, in states such as Ohio, are no longer permitted to allow intermediate algebra to satisfy the requirements of any kind of degree. However, states such as California continued to allow intermediate algebra to satisfy the two-year degree.

Complete College America

Complete College America (CCA) is a nonprofit organization that was established in 2009. Funded by the Bill and Melinda Gates Foundation, CCA's purpose is to increase the number of Americans with college degrees or certificates that can lead to employment (Complete College America, n.d.). CCA became a powerful organization very quickly. In addition to becoming an influential organization within higher education, CCA reported revenue of almost 5 million dollars during its inaugural year (Scherer & Anson, 2014).

CCA became another organization to view developmental education as a barrier to student success. In 2012, under the leadership of president and founder Stan Jones, CCA published the report, *Remediation: Higher Education's Bridge to Nowhere*. In this report, when referring to developmental education, Mr. Jones asserted that "colleges and universities have a responsibility to fix the broken remedial system that stops so many from succeeding" (p. 9).

Bridge to Nowhere was controversial. Advocates of developmental education, such as Dr. Hunter Boylan, warned the public that taking drastic measures against developmental education could harm students (Fain, 2012). Nonetheless, CCA representatives began to approach state legislatures and call to their attention the abysmal success rates in developmental education and its alleged negative impact on higher education (Mangan, 2013). CCA's attack on developmental education intensified the national push to compress, reduce, and even eliminate standalone developmental education.

The Move from Quarters to Semesters and the Compression that Followed

In the 2000s, many community colleges felt pressured to switch from the quarter system to the semester system. Again, semesters are typically 15 or 16 weeks in length, whereas quarters are generally 10 or 11 weeks long. This pressure came from legislatures, state board of regents, and even accrediting bodies in states where some public colleges used the semester system and others used the quarter system. Consequently, Sisco, Bordi, Telford, and Habyan Community Colleges all switched from the quarter system to the semester system. While there are no participants from Lester Community College (LCC) to represent this decade, it is noteworthy that LCC made this change as well.

The Pressure to Compress

Developmental math came under fire in the aughts as data continued to emerge that reflected poorly on the discipline of developmental math. One such finding, which garnered a lot of attention, was that only 33% of students who place into a developmental math course finish their math requirements within 3 years. Another finding that further marred developmental math was that just 17% of those students successfully finish a developmental math sequence of at least three courses (Bailey et al., 2010). Furthermore, students who place into the lowest level developmental class have the lowest chance of completing the sequence (Bahr, 2013; Xu & Dadgar, 2018). Ultimately, many legislatures and administrators felt that developmental math sequences were simply too long and consisted of too many required classes (Edgecombe, 2011). State legislatures considered the overall cost of developmental education too expensive. In 2010, the states' higher education executive officers testified that the cost for developmental education was as high as $3 billion across the United States.

Acceleration allows students to move through their developmental course sequence at a quicker pace, while compression involves shortening that sequence. With these tools, students would still study the same material but would be required to take fewer developmental math classes. Administrators wanted fewer courses, but they also asked for fewer breakpoints (where one course ends and another begins). Professor Sutcliffe elaborated:

> All of a sudden, our administration was worried about too many exit points. They were afraid that if there were too many classes to get through to get to college algebra, we would lose students.

Professor Moyer concurred:

> Okay, so if a student tested into our lowest developmental math class [computational skills], they would have to complete four classes before even getting to college algebra. And only one in 115 students who tested into computational skills even got to college algebra. That's less than 1%. So, our administration wanted a shorter pathway.

This line of thinking never sat well with Professor Mesa, though:

> I would hear administrators complain about the high percentage of students who tested into our computational skills class but did not get into college algebra, and I was like, "duh!" They place into computational skills because they are bad at arithmetic. They have trouble with fractions, decimals, and maybe even multiplying and dividing [whole numbers]. They need a lot of work to prepare for college algebra.

Professor Morgan noticed a shift in priorities:

> They [the administrators] were still worried about success rates, but it became more about persistence and retention. Even if a course had decent success rates, they were more concerned about losing students. For example, a chunk of students would pass our introduction to algebra class but never register for the elementary algebra class. So, it also became about, how do we keep the students even if they passed our classes?

As many community colleges converted from quarters to semesters, the faculty, urged by the administration, attempted to compress their course sequences.

SCC

SCC officially converted to semesters in 2013. Figures 6.1 and 6.2 show the sequence of developmental math courses on the quarter system prior to their compression and the result after the conversion. It is noteworthy that 1 semester credit hour is equal to 1.5 quarter credit hours.

Professor DeSilva elaborated:

> Our dean was really worried that when we went to semesters, it would take even more time for students to get to a college-level math class, so he wanted to make sure it would take shorter, not longer, to get to college-level math.

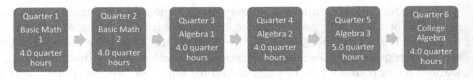

FIGURE 6.1
SCC prior to compression (quarter system).

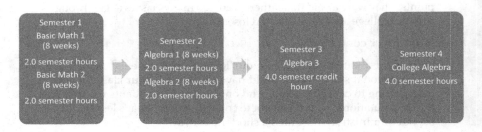

FIGURE 6.2
SCC compression for semesters.

Professor DeSilva explained that SCC implemented "minimesters," something that other schools were doing:

> Minimesters were eight-weeks long so students could complete two math classes in one semester. Students could complete basic math 1 and 2 in one semester. They could complete algebra 1 and 2 in one semester, but we kept algebra 3 [intermediate algebra] as a full semester, so it was a little less time.

How did this impact success rates?

> They stayed about the same. Our success rates for the arithmetic classes were in the low to mid-50s [percent] and the success rates for the algebra classes were in the mid-60s.

SCC also implemented 1-week bootcamps for the five developmental math classes. Students could complete these programs during semester breaks, typically over winter vacation, spring break, or immediately before the fall semester. Bootcamps would meet over 4 or 5 days for a total of 20 hours, and they would cover the entire course content. For example, if a student placed into Basic Math 2, they could complete the bootcamp and begin Algebra 1 at the start of the semester. Successfully completing a bootcamp could shave off either half a semester or a full semester of developmental math. Professor Douglass reflected on the bootcamps:

> These [the boot camps] work well for the lower-level classes for the higher-end students. So, if a student has knowledge of Algebra 1 content but needs a brush-up, this is a great option. They get an intense review and save some time.

However, Professor Douglass noted two mistakes that SCC made with developmental math bootcamps:

> First of all, they [SCC's administration] started letting every student enroll in a bootcamp. Again, these are for the high-end students only. The lower-level students can't take the intensity of the bootcamps because they lack the skillset. So, we had low-level students who would get frustrated because they couldn't keep up.
>
> The other mistake was we [the math department] created bootcamps for elementary and intermediate algebra. Those classes are too difficult to teach in one week, especially intermediate algebra. We had situations where teachers just couldn't get through the content in a week. Quadratic equations and complex numbers are hard, and you just can't rush through them.

BCC

BCC utilized minimesters as well, compressing their developmental course sequence so that a student could place into the lowest level developmental math class and still complete their sequence in 1 year. Figures 6.3 and 6.4 compare the required course sequence on the quarter system prior to the compression to that on semesters following the compression.

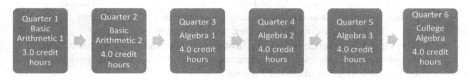

FIGURE 6.3
BCC prior to compression (quarter system).

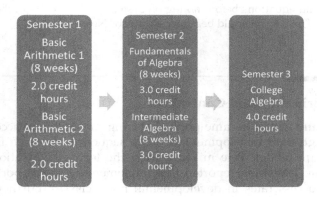

FIGURE 6.4
BCC compression for semesters.

Professor Guzman explained the outcome:

> Our course success rates went down. The success rates for the arithmetic classes stayed about the same. That was about 55%, but the students got killed in the elementary algebra class. The [success] rates went from like 60% to like 45%. It was just too much content for the students to master in such a short time.

Professor Guzman elaborated:

> Students had to go all the way from learning signed numbers to quadratic equations and parabolas in eight weeks. That's hard, but this is what our dean wanted, because they could get through developmental math in a shorter time. With less break points, they retained more students.

TCC

TCC's administration was unhappy with the fact that students who tested into basic arithmetic needed to complete four quarters of math to reach college algebra or statistics. Consequently, TCC's division dean and provost wanted the department to limit the developmental math sequence to 1 year (Figures 6.5 and 6.6).

> We were reluctant to cut down on the arithmetic. Our students were just so weak in arithmetic. Not just fractions and decimals, some [students] had no number sense whatsoever. You could ask a student eight times what is 32, and they couldn't answer, shared Professor Moyer.
>
> The algebra sequence fell nicely. It was eight weeks for the introduction to algebra, eight weeks for the elementary algebra, and eight weeks for intermediate algebra, recalled Professor Mesa.
>
> Even though we compressed [the sequence], it was nice to have so many breakpoints where students would review and take a final exam. They had to master concepts like signed numbers, evaluating expressions, and equations before taking on systems of equations in the next course. The way it should be, stated Professor Noles.

The Emporium Model Continues

The emporium model became popular during the aughts to accelerate students through their developmental course sequence. Research findings on the emporium model were mixed during the teens. The National Center for Academic Transition reported that implementing the emporium model improved success rates in developmental math classes. Furthermore, this model allows students to obtain more individualized assistance and focus more on troublesome topics. It also allows for more time on task as opposed to listening to lectures (Twigg, 2011).

FIGURE 6.5
TCC prior to compression (quarter system).

FIGURE 6.6
TCC compression for semesters.

Additional findings emerged that challenged the emporium model. Some studies noted that students who attempted the emporium model were less successful than those who took math in a traditional setting and were even less likely to persist in college (Childers & Lu, 2017; Kozakowski, 2019). Other drawbacks to the emporium model include a lack of depth: students can get the correct answers by guessing and checking rather than by truly understanding the content (Ariovich & Walker, 2014; Beamer, 2020). Other studies have found little evidence that the emporium model is superior to the traditional approach (Weiss & Headlam, 2018).

BCC continued to employ the emporium model into the teens.

> I feel like it [the emporium model] got better over time. Students knew what they were signing up for. As new instructors came in, they were more open to teaching it. Advisors also understood the emporium model better and that helped our students. I think as we got more experienced, we started anticipating some of the technical and logistical problems that would happen, reflected Professor Mussina.

Professor Guzman elaborated on improvements to the emporium model:

> We would call students one or two days before classes began to explain to them the way the course would work and ask if they needed anything. They really appreciated it, and I think it helped ease them into class. We also decided to have weekly mini-conferences with each of our students. It was a way to check in with them if they had any questions.

I inquired whether BCC requires all developmental math students to use the emporium model or if some classes are still face to face.

> It's about 50–50. Our dean and administration wanted all our dev classes to be in the emporium model, but that just wasn't fair to our students. Some [students] just don't learn math that way. They need guided practice before they start practicing math problems, said Professor Lopez.

What about the success rates in the emporium model?

> They have been about the same as our traditional classes. Sometimes the emporium model's success rates are better; other times the traditional model is better but not by much, shared Professor Guzman.

GCC employed a hybrid emporium model during the aughts and ultimately abandoned it. However, they brought it back in 2013 for all developmental math classes.

> Our administration wanted us to employ the emporium model. It was part of a grant from the Gates Foundation, but they allowed us to do it however we wanted, which was nice. The first thing we did was meet as a department and discuss what went wrong the first time we did the emporium model. Our biggest issue was in the hybrid model; there was no consistency. Some teachers were lecturing most of the time; some were lecturing some of the time, and some none of the time, stated Professor Sutcliffe.

The solution was more structure:

> We decided to make each dev math class part lecture and part lab. Our developmental math classes all met for four hours per week. So, we designed our classes as part lecture and part lab. Each class would be in a traditional lecture format for two hours a week, like on a Monday, and then two hours in the lab on a Wednesday.

Professor Holton described the structure of the course.

> We would introduce the new topics the first time we met during the week (e.g., Monday). There wasn't a whole lot of practice. We would just give them sample problems. Then, later in the week (e.g., Wednesday), we would meet in the computer lab, and they would practice. They could still work ahead if they wanted to, and some did. Students take their exams in the traditional class, but if they worked ahead, they could take their exams in the college's testing center ahead of time.

This model has some benefits and some challenges:

> I like how the students get the best of both worlds. They get the guided practice from us and the drill and practice in that order, and that is the way it's supposed to be. The challenge is we give them so much information when we see them. For example, in one class, I gave short lectures on all of factoring. That's something that usually takes two full classes, because I can introduce a problem and let them practice, introduce a problem and let them practice. With the emporium model, it's so much for them to take in; then, they have to wait two days to do all the practice, stated Professor Thurmond.

Has the emporium model increased success rates in developmental math classes?

> They actually increased a little bit. They went up about five percentage points or so. I think it's just having more structure to the course and having that blend of lecture and drill and practice, reflected Professor Sutcliffe.

As part of a grant from the Gates Foundation, Habyan Community College implemented the emporium model for all their developmental math classes in 2012. Classes were all self-paced labs with no lecture. Professor Bell explained:

> It was a complete nightmare. Most of the students hated it; they couldn't understand why it was a math class and we weren't explaining the material to them first. They were just downright frustrated, shared Professor Bell.
>
> I just felt like we were never getting any teaching done. Students would start the modules on the computer and get lost right away. By the time we walked around to help them, they were already behind and frustrated. After a while, it felt like we were just giving them the answers to put in the computer, stated Professor Hickey.
>
> The emporium model was a mess, but it was the absolute worst for the intermediate algebra students. The material is just so sophisticated. It's very hard for students to teach themselves topics like quadratic equations and complex numbers. We need to give students guided practice before they practice these type of problems, said Professor Williamson.
>
> Our success rates plummeted, *said Professor Bell*. Most of our success rates for the developmental math classes were in the 40s [percent]. By 2016, we abandoned the emporium model and that was that.

Professor Johnson, from Kilgus Community College, conducted program reviews for several community colleges and offered the following opinion on the emporium model:

> The straight emporium model [with no lecture] was pretty much a dismal failure. It was just drill and practice, and there really wasn't any learning going on. It [the emporium model] started because a president or someone went to a conference and discovered you could cut costs by having students work on computers and have fewer faculty. In many cases, the math departments weren't even consulted.

The Decline of Arithmetic Classes

Throughout the aughts and into the teens, the demand for arithmetic classes soared. However, this trend began to change during the teens. The U.S. Department of Education began to restrict federal financial aid for students taking classes with content below ninth grade level (Federal Student Aid, 2021). This decision impacted classes with arithmetic content (fractions, decimals, percentages, proportions, etc.). Community colleges could still offer these classes; however, students would have to self-pay. Problems emerged as schools found that most students who were in arithmetic classes depended on federal aid.

> When we heard about this mandate, the first thing we asked was, "how many of our students who are in these classes are on financial aid," *recalled Professor Guzman*. It was almost 70% of the students who were in our Basic Arithmetic 1 and Arithmetic 2 classes.
>
> More than two-thirds of the students who were in our basic math class were on financial aid, so yeah, that caused us some concern, shared Professor Bell.

How did the community colleges respond?

> At first, we decided to just run the classes and see what would happen for our Basic Mathematics 1 and 2 classes, but most of the classes wound up being cancelled due to low enrollment. Our dean completely freaked out because of all the students we were losing, said Professor DeSilva.

With many students unable to self-pay, community colleges had to find ways to help students who placed into arithmetic classes.

> Here was the problem: we had so many students with really bad arithmetic skills, and they needed those arithmetic skills to be successful in pre-algebra. You can't just go into algebra with no arithmetic background, asserted Professor Mesa.
>
> We decided to reach out to our local adult basic education program. It's state funded and free for students. They agreed to work with the students who tested into our arithmetic classes, recalled Professor Moyer.

Habyan Community College followed the same strategy:

> We were all excited about it. It seemed like a great idea for everyone. Students who needed arithmetic instruction would get free instruction. They could prepare for our placement test and be prepared for our introduction to algebra class. What could go wrong? asked *Professor Hickey*

Unfortunately, the faculty and administration received an answer to that question.

> There were two major problems, *reflected Professor Hickey*. The first was that our students weren't interested in the ABE [adult basic education] classes. See, they don't get financial aid for the ABE classes, because they aren't part of Habyan. So, they looked at it as they would have to take this class for free. So, they just passed on coming to Habyan. The second problem was that our administration didn't support it [students taking classes through ABE]. They were afraid that if we required students to

take ABE classes it would drive them away from Habyan and to another school, so our administration saw it as a money loss.

The reality was the adult basic education classes were very effective. The teachers there were great, and students learned their arithmetic skills. It's just a very unfortunate case of where money prioritizes education, opined Professor Williamson.

GCC attempted to connect their arithmetic students with an ABE program as well, and the results were not good.

There was a school [another community college] about 20 miles from us that was just letting all students into their basic algebra class, so many of our students went there. Needless to say, our administration was not happy, stated Professor Holton.

Our administration wasn't supportive of the adult basic education program from the beginning, *recalled Professor Sutcliffe.* Our dean was afraid that if students took one of those classes [ABE], they would take less classes at Griffin and that would lead to less FTEs for the college.

With the loss in student enrollment, GCC's administration wanted the faculty to allow open enrollment into their Algebra 1 classes.

We fought tooth and nail against it. It was my last fight before I retired, *stated Professor Thurmond.* Our administration just couldn't understand that algebra concepts like linear equations and evaluating expressions are not kindergarten material. These concepts have fundamental prerequisites.

Professor McDonald explained the compromise:

We finally got them to agree to us offering a one-week arithmetic refresher course, kind of like a one-week bootcamp. The refresher courses took place the week before classes started and met every day for about four or five hours. If students passed the course, they could take Algebra 1. They had to self-pay a small amount, and we even used some internal scholarship money to help them.

TCC began employing a similar model to GCC. Professor Noles elaborated:

We just asked ourselves, "what are the most basic arithmetic concepts that students need for our basic concepts [of algebra] class?" Those concepts were fractions, decimals, order of operations, and percentages.

What about the students with such low arithmetic skills where 1 week is not enough time?

For those students, we recommended the adult basic education program. We even helped our students register, but for students who can't multiply or divide or even add or subtract, a one-week course isn't the answer. Heck, I don't even think those [full length] arithmetic classes we used to offer were enough for them, shared Professor Mesa.

BCC began offering an arithmetic refresher course, but unlike TCC and GCC's model, which relied on face-to-face instruction, this model was computer-based. Professor Guzman explained:

> It takes place the week before classes start, and students can work on things like fractions, decimals, percentages, word problems, and signed numbers. We use the ALEKS software, which is computer adaptive, so students are able to focus on arithmetic concepts that they really need. At the end of the week, they take an exit exam, and if they get an 80%, they can register for Algebra 1.

Professor Guzman detailed the benefits of the flexibility of the computerized arithmetic refresher:

> If students don't pass the exit exam, they can continue to use our math lab and practice their arithmetic, as long as they are taking other classes at Bordi. This works great for the students who need more than a week for the arithmetic refresher but maybe don't need to go to adult basic education.

How have the enrollment and success rates been for the arithmetic refresher?

> Our enrollment for those classes isn't anywhere near the enrollment for the traditional arithmetic classes, and our success rates for that class are about 70%, reported Professor Noles.

Several faculty also like the departmentalized exit assessment for the arithmetic refresher:

> We like it because we make up the types of questions that we know students need for Algebra 1, asserted Professor DeSilva
>
> We had some discussions about whether we should use our own exit assessment or make students retake the placement test at the end of the arithmetic refresher, *shared Professor Moyer*. Don't get me wrong; we use ACCUPLACER, and we like it, but because it's computer adaptive, it's not always 100% accurate. When it comes to making sure that students were ready for algebra, we wanted an assessment that had what we wanted.

What about all the students who used to place into arithmetic classes? Where did they go?

> Many of them went toward our business and allied health math classes, stated Professor Desilva.

These classes satisfy 2-year degrees in the business and allied health fields respectively. Because they are contextualized classes, states do not consider them to be remedial classes; however, the content of these classes is not rigorous enough to qualify for states' Transfer Modules.

> The enrollment for those classes [business and allied health math] exploded. It used to be students had to pass the arithmetic classes to get into either of those classes, but our dean took away that requirement, so there is no cut score for those classes, shared Professor Holton.

Still, some faculty thought the change in financial aid requirements, which decimated the standalone arithmetic classes, changed the incoming student population. Professor Bell explained:

> We still get a lot, and I mean a lot, of underprepared students in the business math and the allied health [math] classes, and they have very low math skills, but it's different than the basic math class. I feel like the students in the business and the allied health math class are more focused, because they have their minds set on a business degree or a degree in allied health.

Professor Mesa concurred and elaborated:

> I always felt that many of the students who were in our computational skills class had no direction. They came to community college because someone told them to, but some of them had no idea why they were here, and some didn't even want to be here. To be honest, I think many students took advantage of the financial aid system.

SCC and TCC found a middle ground between the 1-week arithmetic refresher course and the ABE class. Both schools offer individualized tutoring for students in their tutorial centers for students who test into arithmetic. Professor DeSilva clarified:

> If students need or prefer more than a week's review of arithmetic, we set them up in our tutorial center. We have a pre-assessment test that we give them. Then, they work on their weaknesses, say fractions or decimals. We have videos for them and worksheets as well. We basically design an individualized program for them. They meet with a tutor a couple of time a week, but they can also get help when needed. They can come as often as they need [to the tutorial center].

Professor Douglass added:

> When they finish the program, we give them a post-test. Then, they can retake the placement exam to see if they test into algebra.

Is this program suitable for all students who place into arithmetic?

> Not the very, very, very low-level arithmetic students, *replied Professor Trombley*. I mean, if they need a brush-up on division, that's ok, but not if they can't add or subtract, or if they don't know their multiplication facts. If that's the case, they need to take the adult education class.

Professor Trombley noted a shortcoming to this program:

> It really only works well during the summer when enrollment's down. During the fall and spring semesters, when our enrollment is much higher, we just don't have the manpower to serve those students who need individual help in arithmetic. Our tutorial centers get so crowded then.

The U.S. Department of Education never specified an official reason for this change in financial aid rules. However, cost seems to be a likely cause. Many students tested into arithmetic classes, and not many completed them; consequently, a large amount of money went toward failure.

The Lowering of Standards in Higher-Level Math Classes?

Through the years, the curriculum in higher-level math classes, such as calculus, remained stagnant. Unlike the gatekeeper courses, these classes were not bombarded with change initiatives. While some instructors utilized occasional collaborative learning, these classes remained primarily lecture-based due to the fast-pace and rigorous content. Professor Trombley elaborated:

> I try to make my calculus and differential equations classes as interactive as possible. I call on students and I have them share answers with each other, but there is so much to cover. It just requires so much instruction on my part.

In the changing times of community college math, several faculty members felt that standards were dropping in these classes.

> I taught calculus for many years, and by the 2010s, we no longer required students to prove rules and theorems on exams, asserted Professor Bell.
>
> We used to require students to prove the Squeeze Theorem, Rolle's Theorem, the Mean Value Theorem, and L'Hopital's Rule. We still showed the students how to do this in class, but they aren't tested on it, *reflected Professor Timlin*. Students know this and they don't pay attention when we show the proofs.
>
> We didn't change the content, but not requiring students to show the proofs felt like we lost some of the rigor, stated Professor Mitchell.

Why did this change?

> It happened gradually, *reflected Professor Moyer*. It had a lot to do with so many engineering and science students taking calculus. They needed calculus for their majors, but because they weren't math majors, many people felt they didn't need the proofs.

Professor Lopez expressed concern over de-emphasizing proofs in calculus:

> I get it; many students don't need to understand proofs, but what about students who are going to become math education or even math majors? By the time they get to courses like analysis and abstract algebra, students need to understand how to construct a rigorous proof. I can't help think we are doing them a disservice.

Accreditation and Faculty Credentialing Issues

In the four decades leading up to the teens, there was some ambiguity regarding the academic credentials needed to become a community college faculty member. In general, for math, most schools required faculty to possess at least a master's in math to teach. However, unlike the K–12 sector, which generally has more standardized requirements set by the individual state, the hiring standards were set by the individual community college. This changed in the teens.

During the teens, various accrediting bodies (e.g., the Higher Learning Commission) for community colleges began setting requirements for faculty employment. In general, faculty who teach college-level classes at a community college need to possess, at minimum, a master's degree with at least 18 graduate hours in that discipline (Smith, 2015).

Developmental math is not considered college-level math; therefore, many community colleges did not require an extensive math background for hire.

> When I was first hired, developmental education was all about teaching. It wasn't about how much math I knew; it was about how well I could teach the math that was required, shared Professor Thurmond.
>
> I had trouble looking for teaching positions in community college math. It seemed most of them wanted a master's in math. I got the feeling my resume never made it through the front door in those places, but with developmental math, they were more focused on my background. I had taught both elementary and high school, and that, combined with my degrees in education, seemed to make me a good fit, reflected Professor Sutcliffe.

However, this mandate concerned faculty who lacked the mandatory credentials to teach college-level math:

> I was on the verge of retirement when this mandate came down, *recalled Professor Thurmond*, so I didn't worry, but a lot of the faculty who were teaching dev classes, who didn't have the upper-level math, were worried they would get squeezed out of the job.
>
> I never taught the college-level math classes, so it really didn't change things for me that much. It was just freaky how the basic arithmetic class just died, and that was one less class I could teach. It was also the fact that I knew our administration wanted to minimize or even eliminate developmental education. If that ever happened, what would happen to me? questioned Professor Mesa.

The depletion of the arithmetic classes meant the merging of developmental and college-level math for TCC.

> It was a scary time. When computational skills went away, we lost almost half our enrollment [in developmental math]. I was afraid that when we merged departments they would cut the faculty who didn't have the credentials to teach the college-math courses, reflected Professor Mesa.

Fortunately, that was not the case.

> We kept our jobs. Those of us who were teaching developmental math continued to teach the algebra classes. We also teach the business and the allied health math classes. Since those classes aren't in the Transfer Module, you don't have to have the credentials, shared Professor Mesa.

Still, the elimination of the arithmetic courses, combined with the administration's desire to push students through to college-level math, left some uncredentialed faculty uneasy. Consequently, some decided to try and become credentialed by enrolling in online graduate-level math classes. This was the case for Professors Holton and Sutcliffe.

> I figured the only way I could get job security is if I got credentialed. I didn't have the time to take [in-person] classes at my local university, so I looked online to see if I could get credentialed. I was amazed. You wouldn't believe how many universities offer online graduate math certificates just to get community college math professors credentialed. I guess there's truly a market for everything. So, I applied to Price University. All they wanted was for me to have taken Calculus 2. That was it, *recalled Professor Holton*. My first class was abstract algebra.

> I couldn't believe I was accepted at Aldrige University, *exclaimed Professor Sutcliffe*. I had such a weak background. I took three semesters of calculus, differential equations, linear algebra, and that was it. The first class I had to take was introduction to analysis. I figured if they accepted me, I would be fine.

These courses were not a good experience for either faculty member.

> At first it wasn't too bad, *shared Professor Sutcliffe*. We reviewed logic proofs and then functions, but then we got into actually writing a proof, and I was lost. We had these online study groups, but it seemed like everyone else was so far ahead of me and getting it. I just didn't get the basics of how to establish a proof. We were supposed to rely on theorems from calculus, and it had been so long since I took calculus. Heck, when I took calculus, I didn't even understand the theorems. I just did enough to pass. I didn't stick around long enough to take the first test.

> I was kind of ticked off, *said Professor Holton*. They [the admissions staff at Price University] thought I could do this kind of math? The highest math class I took was differential equations. How did they think I could go from that to abstract algebra at the graduate level? I was a stranger in a strange land. I guess, if anything, I needed a very slow-moving class on proof writing, like how to write a proof, because I didn't get the first thing about writing a proof, and I was lost. I guess I finally understood how so many of my students feel.

Both instructors discontinued this plan.

> I'm fine continuing to teach developmental math classes, but with so much uncertainty in the field, it still makes me nervous, stated Professor Holton.

Dual Enrollment

During the teens, dual enrollment became a popular practice in community colleges. Dual enrollment refers to high school students enrolling in college-level classes and earning college credit while in high school.

The concept of dual enrollment can be traced back to the middle of the twentieth century. The University of Connecticut began a partnership with local high schools, as high school officials were concerned that senior year was not challenging enough for students. By the end of the twentieth century, dual enrollment was being practiced in 47 states (Education Commission of the States, 2001). Dual enrollment surged during the aughts, as the number of students enrolled in dual enrollment grew from 1.16 million during the 2002–2003 academic year to 2.04 million during the 2010–2011 academic year (Thomas et al., 2013). The practice continued to grow during the teens as well (An & Taylor, 2019; Flynn, 2019).

Dual enrollment has many advantages. High schools have long sought to increase graduation rates, and colleges have searched for ways to increase enrollment. Dual enrollment provides more incentive for high school students and increases enrollment for colleges (Cassidy et al., 2010). It is also financially beneficial to students, as the costs of dual enrollment are covered by states.

Dual enrollment has been largely beneficial to students. Based on a national study, 88% of the community college students who pursued dual enrollment remained in community college after high school (Fink et al., 2017). The American Educational Research Association reported that students in a dual enrollment class were 9% more likely to earn a bachelor's degree than those who did not (Blankenberger et al., 2017). Additionally, dual enrollment allows students to develop better academic and motivational skills. College classes are more rapidly paced than high school classes, and dual enrollment can prepare students better for college (An & Taylor, 2019).

Dual enrollment has presented some challenges, which include scheduling issues between high schools and colleges. More specifically, it can be difficult to schedule a dual enrollment class that accommodates both the high school and the community college as well as the students' schedules. Dual enrollment grades are also permanent. That is, students who fail a dual enrollment class have their grades permanently transcripted, and reversing this can be challenging (Flynn, 2019).

The faculty at the participating schools have largely positive experiences with dual enrollment.

> Some of the high school students in my college algebra classes struggle at first. They're not used to the fast pacing of the course and the rigor of a college class, but many of them step up to the challenge, shared Professor Williamson.
>
> I've had students who were in my classes when they were in high school [dual enrollment] come back and thank me for working with them and showing them what college life was like, reflected Professor Guzman.

Professor Bell agreed:

> There have been students who said they were afraid of coming to college, so they decided to try some classes their senior year. They said I provided such a calming influence on them and showed them they really could do math in college.

For Professor DeSilva, dual enrollment is more personal:

> As a former high school teacher and a parent, I see and have seen some of the trouble high school kids get into, especially the kids who really don't have any direction in their lives. Kids who enroll in dual enrollment see a purpose and work toward something after high school. That's important.

The faculty shared some challenges they have encountered with dual enrollment:

> This has been a complaint of mine for years. I don't think math is as rigorous in high school as it needs to be, *shared Professor Moyer.* For college algebra and trig[onometry], students need a solid and deep conceptual understanding of algebra and some geometry. I don't always feel they have that, and they struggle in my college algebra class.

Professor Timlin discussed the grading challenges:

> The grading is a big issue. We [at GCC] like at least 80% of our final grade to be based on proctored exams or quizzes. The high schools like to allot so much of the grade for participation and attendance. First of all, that explains why we get underprepared students in our classes. Second, it causes a lot of conflict in dual enrollment. It seems like we are always fighting with the high schools over the grade configuration.

Professor DeSilva surfaced the issue between excused and unexcused absences:

> In high school, students can be absent and basically be excused from their assignments if their parents write a note. I can't tell you how many notes from parents I have received. I have to explain that is not case in college. It doesn't matter why you were absent. You need to do the work and be ready for the test.

In summation dual enrollment has positively impacted community college math students. Additionally, dual enrollment has allowed William Rainey Harper's vision of students completing general education college courses at the high school level to come into fruition.

Multiple Measures

The validity of computer adaptive placement exams has been debated since their inception. Some researchers have determined placement exams, such as ACCUPLACER, to be reliable instruments (James, 2006; Donovan & Whelan, 2008). However, other researchers such as Clayton et al. (2014) found that one in four students who took a math placement test was placed incorrectly, and most of the time these students needed to unnecessarily complete additional developmental math classes. Consequently, in the teens, community college administrators began seeking alternative methods for placement.

To increase enrollment in college-level courses and to ensure better overall student placement, Long Beach City College (LBCC) in California became a pioneer in the use of multiple measures. In 2012, LBCC considered high school data such as a student's grade point average (GPA) as well as ACT and SAT scores for placement. More specifically, if students scored high enough on their ACT or SAT scores and had a high enough GPA, they could enroll in a college-level class. Consequently, 30% of incoming students placed into a college-level math class, as opposed to only 9% from the previous year (LBCC, n.d).

As the teens progressed, the use of multiple measures spread nationally. Several states, such California, Texas, Nevada, Oklahoma, and Colorado either allowed or required the use of multiple measures for placement (Ganga & Mazzariello, 2019). While many more students are placing into college-level math, it was unclear, by the end of the teens, as to whether students who placed into college-level math using multiple measures were more success-ful in such classes than those who placed in using a traditional placement exam. Additionally, schools are struggling with the correct algorithm (com-bination of placement score, high school GPA, ACT and SAT scores) for the best predictor of proper placement (Goudas, 2019; Barnett, Kopko, Cullinan, & Belfield, 2020).

What has been the faculty experience utilizing multiple measures? Telford Community College implemented multiple measures, and in addition to a student's placement score, the school considers SAT and ACT scores. Are more students testing into college-level math? Are students less prepared for college-level math?

> I don't think it's made too much of a difference, *shared Professor Noles*. I check the prerequisites of all my students in all my classes before the semester starts. Students who place into my math classes via the SAT or ACT scores are doing just fine. It helps a little more of our students place into college-level math, and they are doing all right, but here is the thing to remember. I have found the lower-level math students scored very low on their ACT exams or didn't even take them, so if anything, this [mul-tiple measures] helps the higher-end students who may have had a bad day on the placement exam.

Sisco Community College has also employed multiple measures by employ-ing ACT and SAT scores, and they have had serious discussions regarding using a student's GPA as well.

> I have a lot of concerns about high school GPA being used for math place-ment, *asserted Professor DeSilva*. First of all, it's never been made clear whether it's their math GPA or their overall GPA. So, if a student gets an A in music, art, or even gym class, that could impact where they start math in college?

Professor Trombley expressed similar concerns:

> I get that a strong GPA may indicate a good student, but being success-
> ful in math is not just about being a good student. You need the proper
> prerequisite skills in math.

Professor Douglass expressed another concern regarding the use of a high
school GPA:

> Remember, some teachers give warm body points. Some students get
> way too many points simply for showing up to class. Even if the GPA
> reflected their studies in math, how do we know how much of the grade
> focused on exams and quizzes and knowing the content?

Common Core Forms Below

Researchers continued to search for ways to improve K–12 education, align
standards, and increase college readiness. In 2009, the National Governors
Association assigned several renowned educators and researchers to develop
the standards for Common Core, which are national standards that focus on
English and mathematics. Regarding mathematics, Common Core emphasizes
benchmark standards for each grade level so that students are better prepared.

Common Core places an emphasis on modeling and more abstract learning as opposed to drill and practice. The Common Core math standards
also encourage diving deeper into topics such as problem-solving skills,
operations with whole numbers, ratios and proportions, and basic algebra,
as opposed to attempting to cover more topics.

On a national level, Common Core math does not dictate pedagogical practices. However, on a state level, teachers, especially at the elementary school
level, are encouraged to teach certain topics in a specific way. In some cases,
students must demonstrate these methods on state exams. Figure 6.7 compares
the traditional method of subtraction and subtraction using Common Core.

Before joining BCC, Professor Guzman was an elementary school teacher
and utilized Common Core math.

> Common Core focuses on the way we think as we do math. So rather than
> have students memorize math facts, they are able to think through the pro-
> cess, which can help develop a deeper conceptual understanding of math.

However, Professor Guzman expressed some concerns:

> I feel that with Common Core there is sometimes too much guessing and
> thinking. We can't just focus on math facts and procedures, but students
> need to learn their math facts, and they need to develop solid organiza-
> tional skills. That is why we see so many of these students in developmen-
> tal math. We just need to find a middle ground between math facts and
> organizational skills and helping students develop deeper thinking.

```
A Traditional Subtraction Problem:

    52
  - 11
    41

Subtraction Using Common Core:

52 – 11= ___

11 + 9 = 20
20 + 10= 30
30 + 10 = 40
40 + 10 = 50
50 + 2 = 52
    41  ←answer
```

FIGURE 6.7
Common core question.

By the late teens, it was inconclusive as to whether Common Core sufficiently prepares students for college. I asked the faculty whether Common Core impacts and will impact future community college students, especially in gatekeeper math courses.

> A big reason students place into developmental math and come to college unprepared is that they develop gaps in their math, and over the years the gaps widen, *shared Professor Timlin*. I don't know if different ways of teaching or having these standards will help. What students need is better intervention over the years in elementary and high school to narrow those gaps.

Professor McDonald shared some of the concerns as Professor Guzman.

> Call me cynical, but we've seen this before, a firehose approach to fix math. I like how Common Core tries to develop more critical thinking, but it goes too far. Students need to develop math facts, procedures, and yes develop memorization skills. That's how they develop number sense. I wish we would stop going from one extreme to another.

Summary

The aughts transitioned into the teens, and changes to introductory community college math classes continued. Community colleges were pressured to compress their developmental math course sequence so that students who placed into developmental math reached their college-level math classes at a quicker pace. Additionally, community college math programs continued to employ the emporium model to try and accelerate students and improve success rates with mixed results. The landscape of community college math began to change as well. The U.S. Department of Higher Education restricted federal financial aid for college classes that contained content below the ninth-grade level. Consequently, this spelled the end of quarter or semester long standalone arithmetic classes. As the teens progressed, more changes would surface for community college math.

References

An, B., & Taylor, J. (2019). A review of empirical studies on dual enrollment: Assessing educational outcomes. *Higher education: Handbook of research and theory.* doi: 10.1007/978-3-030-03457-3_3.

Ariovich, L., & Walker, S. A. (2014). Assessing course redesign: The case of developmental Math. *Research & Practice in Assessment, 49,* 45–57.

Bahr, P. R. (2013). The deconstructive approach to understanding community college students' pathways and outcomes. *Community College Review, 41*(2), 137–153. doi: 10.1177/0091552113486341.

Bailey, T., Jeong, D. W., & Cho, S. W. (2010). Referral, enrollment, and completion in developmental education sequences in community colleges. *Economics of Education Review, 29*(2), 255–270.

Barnett, E. A., Kopko, E. A., Cullinan, D., & Belfield, C. (2020). Who should take college-level courses? Impact findings from a multiple measures assessment strategy. *Community College Research Center.* https://ccrc.tc.columbia.edu/media/k2/attachments/multiple-measures-assessment-impact-findings.pdf.

Beamer, Z. (2020). Emporium developmental mathematics instruction: Standing at the threshold. *Journal of Developmental Education, 43*(2), 18–25.

Blankenberger, B., Lichtenberger, M., & Witt, M.A. (2017). Dual credit, college type, and enhanced degree attainment. *Educational Researcher, 46*(5), 259–263. doi: 10.3102/0013189X17718796.

Cassidy, L., Keating, K., & Young, V. (2010). *Dual enrollment: Lessons learned on school level implementation.* U.S. Department of Education. Office of Elementary and Secondary Education, Smaller Learning Communities Program. https://www2.ed.gov/programs/slcp/finaldual.pdf.

Childers, A. B., & Lu, L. (2017). Computer based mastery learning in developmental math classrooms. *Journal of Developmental Education, 41*(1), 2–6, 8–9.

Complete College America. (2012). Remediation: Higher education's bridge to nowhere. https://completecollege.org/wp-content/uploads/2017/11/CCA-Remediation-final.pdf.

Complete College America. (n.d.). About. Author. https://completecollege.org/about-us/.

Donovan, W. J., & Wheland, E. R. (2008). Placement tools for developmental mathematics and intermediate algebra. *Journal of Developmental Education, 32*(2), 2–11.

Edgecombe, N. (2011, May). *Accelerating the academic achievement of students referred to developmental education (CCRC working paper no. 30).* New York, NY: Community College Research Center, Teachers College, Columbia University.

Education Commission of the States (2001). *Postsecondary options: Concurrent/dual enrollment.* Denver, CO. http://www.ecs.org.

Fain, P. (2012). Complete college American declares war on remediation. *Inside Higher ED.* https://www.insidehighered.com/news/2012/06/19/complete-college-america-declares-war-remediation.

Federal Student Aid. (2021). *2021-2022 Federal Student Aid Handbook.* https://fsapart-ners.ed.gov/knowledge-center/fsa-handbook/pdf/2021-2022.

Fink, J., Jenkins, D., & Yanaguira, T. (2017). *What happens to students who take commu-nity college "dual enrollment" courses in high school.* Community College Research Center, Teachers College, Columbia University. https://files.eric.ed.gov/fulltext/ED578185.pdf.

Flynn, K. (2019, May 29). Pros and cons of dual enrollment. Saving for college. http://www.savingforcollege.com/article/pros-and-cons-of-dual-enrollment.

Ganga, E. & Mazzariello, A. (2019). Modernizing college course placement by using multi-ple measures. *Center for the Analysis of Postsecondary Readiness.* https://postsecond-aryreadiness.org/modernizing-college-course-placement-multiple-measures/

Goudas, A. M. (2019). *Multiple Measures for College Placement: Good Theory, Poor Imple-mentation.* Community College Data: Sharing, Analyzing, and Interpreting Data in Higher Education, with a Focus on Community Colleges (Working Paper No. 6). http://communitycollegedata.com/articles/multiple-measures-for-college-placement/

James, C. (2006). Accuplacer online: Accurate placement tools for developmental pro-grams? *Journal of Developmental Education, 30*(2), 2–8.

Kozakowski, W. (2019). Moving the classroom to the computer lab: Can online learn-ing with in-person support improve outcomes in community colleges? *Economics of Education Review, 70,* 159–172. doi: 10.1016/j.econedurev.2019.03.0054.

Long Beach City College (n.d.). *Promising pathways to success: Using evidence to dra-matically increase student achievement.* https://www. accca.org/files/Awards/Mertes%20Award%20-%20LBCC%20Ex%20Summary.pdf.

Mangan K. (2013). How gates shapes higher-education policy. *Chronicle of Higher Education,* A24–25.

Scherer, J. L., & Anson, M. L. (2014). *Community colleges and the access effect: Why open admissions suppresses achievement.* New York: Palgrave Macmillan.

Scott-Clayton J, Crosta P. M., & Belfield C. R. (2014). Improving the targeting of treat-ment: Evidence from college remediation. *Educational Evaluation and Policy Analysis, 36*(3), 371–393. doi: 10.3102/0162373713517935.

Smith, A. A. (2015). Questioning teaching qualifications. *Inside Higher Ed.* https://www.insidehighered.com/news/2015/10/20/colleges-and-states-scramble-meet-higher-learning-commissions-faculty-requirements.

State Higher Education Officers. (2010). *State higher education: State higher education finances,* FY 2010. Author.

Thomas, N., Marken, S., Gray, L., & Lewis, L. (2013). *Dual credit and exam-based courses in U.S. public high schools: 2010–2011.* Washington D.C.: National Center for Educational Statistics. https://files.eric.ed.gov/fulltext/ED539697.pdf.

Twigg, C. A. (2011). The math emporium model: Higher education's silver bullet. *Change: The Magazine of Higher Learning, 43*(3), 25–34.

Weiss, M. J., & Headlam, C. (2018). A randomized controlled trial of a modularized, computer- assisted, self-paced approach to developmental math. *Journal of Research on Educational Effectiveness, 12*(3), 484–513. doi: 10.1080/19345747.2019.1631419.

Xu, D., & Dadgar, M. (2018). How effective are community college remedial math courses for students with the lowest math skills? *Community College Review, 46*(1), 62–81. doi: 10.1177/0091552117743789.

7

The Teens Part 2: Alternative Pathways Lead to Signs of Reform

Like the aughts, the teens consisted of reform in the introductory community college math classes. Community college math departments compressed developmental math course sequences and tried various initiatives to improve student success. However, it was not till the latter part of the teens that the structure of introductory community college math truly began to change.

New and Returning Participants

In Chapter 7, Professors Thurmond (Griffin Community College), Bell (Habyan Community College), McDonald (Griffin Community College), Mesa (Telford Community College), Sutcliffe (Griffin Community College), Moyer (Telford Community College), Trombley (Sisco Community College), Lopez (Bordi Community College), Mussina (Bordi Community College), DeSilva (Sisco Community College), Guzman (Bordi Community College), Holton (Griffin Community College), Hickey (Habyan Community College), Williamson (Habyan Community College), Noles (Telford Community College), and Douglass (Sisco Community College) continue to convey their experiences of teaching community college math in the teens. Furthermore, two new faculty members provide their knowledge as well.

Professor Harnich

Professor Harnich began teaching developmental and college-level math full time at Bordi Community College (BCC) in 2014. She has a bachelor's and a master's degree in math.

DOI: 10.1201/9781003287254-7

Professor Smith

Professor Smith began his teaching career at Griffin Community College (GCC) in 2008. He has a bachelor's and a master's degree in math and teaches both developmental and college-level math.

Alternative Math Pathways

College algebra became a popular course in the 1950s and 1960s since the content in college algebra provided a necessary pathway to calculus and other high-level math classes. This was due to Sputnik and America's push to draw more students into the mathematics and science fields. However, in the twenty-first century, community colleges began to question this pathway.

Requiring college algebra for all or even most college students seemed to be a dated requirement. Did topics such as functions, logarithms, and conic sections serve students who were entering non-mathematical fields? Were there better ways to apply math to the real world for non-STEM majors?

In 2009, the Carnegie Foundation for Teaching Advancement launched a $13 million initiative that focused on increasing student success in community college math while also exploring alternative pathways for students (Merseth, 2011). Additionally, Carnegie established the Networked Improvement Community (NIC). The NIC consists of mathematical experts, community college faculty and administrators, and various change agents (Byrk et al., 2015). This group worked to establish the content and layout for two novel math pathways, Statway and Quantway. Both pathways were designed as alternatives to the college algebra pathway for non-STEM students, allowing students who test into developmental math to complete their college-level math requirements within 1 year.

Statway

The traditional Statway consisted of a developmental component, which took place in the first semester, and the college-level statistics course, which took place in the second. The developmental component focused on topics that specifically prepared students for the college-level portion (Figure 7.1).

FIGURE 7.1
Two-semester statistics pathway.

Quantway

Like Statway, Quantway began as a two-semester sequence for non-STEM majors. Many institutions refer to this sequence as Quantway 1 and Quantway 2. The college-level course, quantitative reasoning (QR), consists of real-life concepts, such as data description, probability, finance problems, and linear and exponential modeling. The developmental portion consists of concepts that prepare students for quantitative reasoning.

The Carnegie Foundation and another national organization, the Charles A. Dana Center, were heavily involved in the development of Quantway. Both organizations stressed the development of critical thinking skills through group-based instruction, as opposed to traditional lecture. This emphasis is based on the belief that critical thinking is important to undergraduate education as a whole and to future job skills (Elrod, 2014). As in previous decades, many educators and administrators asserted that group-based instruction and inquiry allows students to develop meaningful connections with each other and a build a sense of community (Clyburn, 2013; Altose, 2018) (Figure 7.2).

FIGURE 7.2
Two-semester quantitative reasoning pathway.

Corequisites

To further accelerate students through their math requirements, many schools began employing corequisites in the mid-to-late teens. When students place into developmental math, they can take a preparatory or booster class simultaneously with their college-level class. For example, if a student needs to take QR but places into developmental math, he or she could take a QR booster alongside the college-level course. This booster course may meet directly before the college-level course, or it may meet on different days. Ultimately, the booster course attempts to prepare the student very quickly for the college-level course. As an example, a student may learn about slope and equation of the line in the booster course and then apply it later that day or the next day to finance questions. Rather than a two-semester sequence of Statway or QR, the corequisite allows the students to complete both the developmental and college-level component in one semester.

The Community College of Baltimore was the first to use corequisite education when they began employing the Accelerated Learning Model in the late 1990s. In this model, students could complete their college-level requirements along with their developmental requirements.

The Statway and Quantway pathways have been successful on a national level. In 2020 Carnegie reported the findings of a study consisting of 410 students across six community colleges. Students who enrolled in either Statway or Quantway were more than three times as likely to complete their math requirements than those who enrolled in a developmental math sequence. Students also completed their college-level math credits in less than a quarter of the time.

In addition to QR and introductory statistics, some community colleges have implemented college algebra with a corequisite component. In general, this model's booster class consists of content from intermediate algebra, which students must assimilate and apply to college algebra concepts. Studies have shown that students who enroll in college algebra with the corequisite do as well as those who attempt college algebra without the booster course. More specifically, the booster course taken with college algebra has proven effective (Vandal, 2016; Smith, 2019).

The Application of Statway and Quantway

Both Sisco Community College (SCC) and BCC began employing the two-semester sequences of Statway and Quantway in 2014 and 2015 respectively. In both cases, Statway 1 and Quantway 1 were the lowest level of developmental math classes.

> I wasn't sure about it [Statway] at first, but I liked how in the remedial sections, we covered topics that students specifically needed for statistics. We focused on mean, median, and mode, the basics of data description, a lot of basic arithmetic, and then concepts like roots and radicals, shared Professor DeSilva.

For so long when I would teach factoring and quadratic equations, students would ask, "When are we going to need this?", and if they didn't need more math, like calculus, I had no answer. The concepts that we covered in the remedial portion of Statway and Quantway related directly to the college-level portion, stated Professor Guzman.

Professor Guzman also enjoyed the real-life applications in the QR course:

I couldn't believe how much fun I had teaching that course. I like algebra and pre-calculus, but I loved having discussions with the students about probability and debt-to-income ratios. It led to some lively class discussions.

After 2 years of implementing Statway and Quantway, both BCC and SCC found that students were completing their math requirements at a quicker pace.

Students were completing Statway and Quantway twice as fast as the traditional algebra sequence, shared Professor DeSilva.

Our school was obtaining more funding because students were moving through their math requirements more, and we were retaining 20–25% more students than before these pathways, stated Professor Guzman.

However, the faculty expressed concerns regarding the student population in Statway and Quantway 1. While both BCC and SCC utilized a 1-week arithmetic refresher course, Professors Harnich and Guzman felt that some Statway and Quantway 1 students were still underprepared:

We were getting students who had such deficient math skills, *shared Professor Guzman*. In both classes, I was trying to teach solving equations, but some students had never seen equations before, and it took them forever to try and understand. It takes time and practice to understand even a two-step algebraic equation, and some students never caught on.

Even before we can prepare them for statistics or quantitative reasoning, they need some math background. They need solid arithmetic skills and some basic algebra. Among other things, I had students who didn't understand how to evaluate expressions, which they need for both quantitative reasoning and statistics. It seems easy when you know how to do it, but it takes time to learn, asserted Professor Harnich.

Using Quantway 1 and Statway 1 as the lowest developmental math course at SCC was an indicator that they needed an arithmetic refresher. Professor Douglass explained:

We were referring low-level students to the adult basic education classes, but it wasn't enough. We were getting students in those classes [Quantway 1 and Statway 1] who didn't understand rounding numbers; they didn't understand converting decimals to percents; they had trouble with multistep problems. Yes, yes, yes, the calculator can help with arithmetic, but there are concepts the calculator can't help with.

As the teens continued, Statway and Quantway became more popular. To further accelerate students, between 2016 and 2018, Griffin, Habyan, and Telford Community Colleges all implemented the corequisite model, and

FIGURE 7.3
Statistics or quantitative reasoning pathway with a corequisite and an introduction to algebra prerequisite.

SCC and BCC converted their two-semester pathway to the one-semester corequisite model. All five schools mandated an 8-week introductory algebra or pre-algebra course for those students who did not test into the corequisite course. Students needed to score high enough on the ACCUPLACER test to enroll in either introduction to statistics or QR with the booster course. Figure 7.3 shows the Quantway or Statway pathways with a booster course.

> We were already concerned enough about students taking the remedial portion at the same time as the college-level course. If students still needed extra preparation, we wanted to make sure they had that, stated Professor Williamson.
>
> Quantitative reasoning and statistics classes are hard, and even with the booster, we wanted to make sure students came in with basic skills, like signed numbers, evaluating expressions, and solving equations, recalled Professor Moyer.

Statway with the Corequisite Has Gone Well

> I don't know why we never did this before. It's a really a good setup; students who need more remedial math can take our introduction to algebra course. Statistics with the booster works really because for most concepts, students can learn them pretty quickly and then apply them, opined Professor Hickey.

Professor Williamson elaborated:

> You need a good arithmetic background and minimal algebra to do basic statistics, so this worked out well. I appreciated the extra time [in the booster course] to simply review and give students extra practice. It was also good that the students in the booster course got extra practice with the graphing calculator. That is where statistics students really struggle, with the syntax of the calculator.

The students appreciated the opportunity to complete their statistics requirements at a quicker pace, and the results showed it. Professor Noles elaborated:

> Students who attempted statistics with the booster course have a 65% success rate. In other words, 65% of them pass the college portion of statistics. Students who attempt the second algebra course only have a 40% chance of staying around and eventually completing college algebra.

Professor DeSilva shared similar data:

> Students who start our statistics with the booster have a 71% success rate, but only 41% of the students who start our elementary algebra [second algebra course] pass our college algebra class.

To clarify Professor Noles' and DeSilva's comments, students at TCC and SCC started in introduction to algebra, if needed. From there, students could either enroll in elementary algebra (second algebra course) or statistics or quantitative reasoning with the booster course. In summation, students were having more success in statistics or QR than elementary algebra.

Still, Professor Bell still sees the advantage of having students complete an entire algebra sequence:

> Statistics is a tough course. There are lots of very long problems, like bell curves and regressions. It takes a great deal of mathematical maturity to be able to do those problems carefully and correctly, and that's the kind of experience and maturity they developed by taking elementary and intermediate algebra, but I get it. Students move through their math requirements quicker, so I guess this [statistics with the booster] is for the better.

The term mathematical maturity is used informally; however, in his book, *Breaking Barriers: Student Success in Community College Mathematics,* Brian Cafarella (2021) asserted that a student with sufficient mathematical maturity understands the prerequisite skillset needed to master new mathematical topics. Furthermore, a student with high mathematical maturity understands the time and organizational skills required to learn new material.

QR with the Corequisite Has Been a Tougher Transition

The schools struggled with the implementation of QR courses with the corequisite model. This was because a pedagogical mandate from the previous decade resurfaced. Professor Hickey explained:

> It wasn't so much the content in the quantitative reasoning course that was the problem; I actually like the content of this course and how it applies to everyday life. It was the way they wanted us to teach this course. They wanted us to use inquiry or group-based instruction. In other words, the students were supposed to learn the material in small groups, and we, as the instructor, acted as a facilitator.

Who are "they?"

> Basically, the Dana Center and Carnegie [Foundation] had a lot of input in this course. They wanted the course [QR] taught using the inquiry method, so they started going to the individual state's board of regents and telling them, "This is how the course should be taught." So, the state would appoint these advisory panels who would dictate to schools how

the course should be taught. So, then our administration would want us to teach the course using inquiry so they wouldn't get into trouble, I guess, stated Professor Williamson.

In the teens, the QR faculty ran into the same problems as the developmental math faculty did in the aughts when utilizing group-based instruction. Professor Noles explained:

> It was so confusing to the students. We had to put students in these groups, and they had to learn the material by talking with each other. That's very hard to do with tough concepts like probability and those finance questions.

Professor Williamson added:

> Math is difficult for a lot of students, and they need good instruction. If students don't know what they're doing, they can get confused, frustrated, and even develop some very bad habits.

The faculty at BCC and SCC struggled with inquiry-based instruction as well.

> Our problem was we were trying to get the course approved in our state's Transfer Module so quantitative reasoning would transfer for students [when they transferred to a four-year school], but our state's committee said they wanted activity-based instruction. So, our administration said we had to use group-based instruction to get the course approved, recalled Professor Guzman.
>
> The most frustrating part was these people who kept pushing the group-based instruction. We kept telling them that our students need some traditional instruction and guidance, but they kept telling us we needed to push the group-work. They just kept telling us to rearrange the groups, and it would eventually work out; it didn't, shared Professor DeSilva.

I asked Professor DeSilva about the people who continued to push inquiry-based instruction:

> Mostly, it was people from our state who oversaw the quantitative reasoning course. Generally, they were the ones who approved the course going into the Transfer Module. It was also faculty and administrators who loved the idea of group-based instruction and believed everyone should do it.

Professor Trombley, who had to use inquiry in the aughts, felt reunited with a past demon:

> I couldn't believe the "experts" were pushing inquiry again. I was like "Oh no, not this nonsense again." The only difference was they were pushing it for a new course. I was excited to teach quantitative reasoning, but I wasn't living through that nightmare again. So instead of doing all the group-work they wanted us to do, I closed my classroom door and taught my students.

The faculty stated that they gradually brought back traditional teaching methods to the QR course:

> I don't care what our administration or the state says, *asserted Professor Moyer*. My job is to ensure that I provide my students with solid instruction, so now I introduce a topic and explain it to my students.
>
> I still use group work, but first I explain things like box plots, probability, and compound interest problems, stated Professor Noles.
>
> It's not like I just lecture to my students. When I explain a new topic, it's like a group discussion. I introduce the problem and together we work through the problem, and I still use group-work, but I let the students work together after I have explained the concept and after I am sure they have developed a basic foundation for the topic. Just let me do my job, said Professor Williamson.
>
> Once we [the faculty] stopped pushing the group-work, the course was great. I loved teaching the real-world applications, like debt-to-income ratios, and showing them how they can use them in real life, recalled Professor DeSilva.

Group-Based Instruction Aside, QR Has Been Beneficial

Like Statway, Quantway allowed students to move through their math requirements quicker.

> We looked at two years of data, and 69% of the students who started in QR with the corequisite completed the course. Only 40% of the students who started in elementary algebra [BCC's second algebra course] finished college algebra, shared Professor Guzman.

This was the case at HCC as well:

> This [QR pathway] is helping our students. 75% of the students who start Quantitative Reasoning with the coreq[uisite] complete the course. 38% of the students who start our [second] algebra class pass college algebra, shared Professor Williamson.

The faculty at TCC enjoyed the real-life applications of the QR course and felt they allowed the students to be more engaged. As discussed in Chapter 4, TCC implemented general mathematics, a course for non-STEM majors. This college-level course was designed for the Transfer Module. I asked Professor Moyer how the general mathematics course differed from QR.

> I think we make more of an effort in the QR course to relate the content to students' lives. There were word problems in the general math class, but these word problems [in the QR course] are stronger and more modern.

Professor Moyer added:

> There is overlap between the general math and the QR class, but the QR class has more relatable content. For example, the general math class had logic and logic proofs, and students didn't see how those related

to their lives at all. But the QR class has debt-to-income ratio problems, which students really like. So, the QR class just has better content and better applications.

A Closer Look at Corequisites: The Good, the Bad, and the Ugly

Statistics and QR courses employing the corequisite model have positively impacted students.

> I like how students can learn material, like slope and equation of the line or equations with fractions, and then immediately apply them to the college-level QR course. They don't have to learn it and then wait a semester to apply it, shared Professor Moyer.
>
> I can see how corequisites help students move through their coursework quicker. As a teacher, it's frustrating for me when students pass a math class, but because of life circumstances, can't take the next math class. That happens too much. So, if students are willing to put the work in, it's good they can get both their remedial and college-level math requirements done in one semester, reflected Professor Williamson.

However, the corequisite model includes a lot of time and work, and it is not a fit for every student.

> Many students don't understand the time and effort that goes into the statistics and QR pathways. Students have to complete the work for the booster class plus learn material and then quickly apply it to the college-level class, shared Professor Hickey.
>
> I get a lot of students who just get burned out about a quarter to midway through the semester. Their attendance in the booster course gets sporadic midway through the semester. With our statistics class with the booster, they are taking math for over five hours per week. Combine that with their other coursework, it's just too much, said Professor Noles.

In some cases, corequisites are a complete failure.

> We have to be really careful with the college algebra and the booster course. It doesn't work well for students who have a very weak intermediate algebra background, opined Professor Mussina.

Professor Noles concurred:

> Intermediate algebra is hard. A lot of people don't get that. Complex numbers, quadratic equations, parabolas, and absolute value equations: those are hard concepts. It's really, really hard to learn quadratic equations in the booster course and then apply them to polynomial functions or logarithms in the college-level course.

Is college algebra with the corequisite a good initiative?

> I think it can work for students who already have a good understanding of intermediate algebra but didn't test into college algebra. They just need a brush-up on some intermediate algebra, shared Professor Guzman.

The Elimination of Standalone Developmental Math

To increase student completion rates even further, some community colleges have dropped standalone developmental education courses altogether and have allowed students to enroll in college-level math classes with a booster course as a form of default placement. Florida set the trend in the early to mid-teens with three components for major developmental education reform. First, developmental math courses became optional. Consequently, more students began enrolling into introductory college-level classes. Additionally, these introductory college-level classes became more heterogeneous as more Black and Hispanic students enrolled in these classes. Second, Florida reformed how they offered developmental math classes. Schools followed the compression and modularized model, which allowed students to complete their developmental course work in less time as well as focus on difficult content. Instructors also contextualized content more to relate to real-life situations. The third component involved providing more advising and tutoring for underprepared students. According to the research, students in Florida are completing their college-level math courses at a quicker rate (Park-Gaghan et al., 2020). In the late teens, Georgia followed in eliminating standalone remedial classes. Georgia has also reported higher completion rates in college-level math (Complete College America, 2021). Additionally, in the late teens, California, Texas, and Tennessee followed suit with similar reform.

Nationally, the elimination of standalone developmental math has been met with faculty resistance for two reasons: Faculty are concerned that students will not be prepared for their college-level classes and will ultimately fail. Second, faculty are concerned that there will not be enough resources to support students who are attempting to complete introductory college-level math without standalone developmental math (Lewis, 2021).

In 2018, both BCC and GCC removed their standalone developmental math classes, including the arithmetic refresher, for students requiring QR and statistics classes. Griffin even eliminated their entire standalone developmental math sequence and utilized college algebra with the corequisite as the default placement course. Both states were moving this way, and, consequently, this initiative was mandated by their administrations. Professors Guzman and Mussina explained how this impacted the QR and introductory statistics courses.

> Don't people understand the kinds of students you get when you eliminate developmental education? I'm trying to explain histograms, and I have students who don't even understand how to come up with the intervals along the x-axis, shared Professor Guzman.

Professor Mussina discussed how this change impacted the flow of the class:

> I'm trying to explain how to do probability or how to do a debt-to-income ratio problem, and I have people who are struggling with very basic arithmetic, and they keep asking me questions. One woman

wouldn't let me move on because she didn't understand rounding whole numbers. It takes me away from teaching the stuff I'm supposed to be teaching.

Professor Harnich described the inequity of the elimination of standalone developmental math:

Granted, there are a lot of low-level students who don't pass introduction to algebra anyway, but I think we are doing such a disservice to the mid-tiered students. These students just need a more solid foundation before attempting college-level math, but we're not giving them the chance.

The participants noted the importance of reading skills in a QR or statistics class. By removing standalone developmental courses, students entered QR and statistics not only with low math skills but also with very low reading skills.

It has been tough, *stated Professor Guzman.* Quantitative reasoning and statistics classes are hard classes that have tough content and a lot of reading. When you just allow anyone to register for those classes, you get people who can barely read.

In the QR course, there are so many word problems, and I have people who can hardly read. I can't help people with probability applications if they can't read the word problems, shared Professor Mussina.

Schools need to make sure that people who register for statistics or quantitative reasoning can read at a sufficient level. We can't teach people how to read and teach them the math skills they need for quantitative reasoning or statistics and teach them the college-level content in those courses in just one semester, asserted Professor Lopez.

Does Eliminating Standalone Developmental Math Increase Completion Rates?

One confusing matter remains: How does the elimination of standalone developmental math yield higher overall completion rates despite creating chaos? Tables 7.1 and 7.2 feature data from GCC, and they explain how this can be true. In the fall of 2017, GCC required students who placed below either QR or introduction to statistics with the booster course to complete introduction to algebra. In the fall of 2018, introduction to statistics and QR with the booster course was the default placement.

Professor Smith elaborated:

The administrators, legislators, and others were looking at the students who started introduction to algebra and didn't pass. When they put them all in college level courses with the coreq, the overall completion rates went up. We retained more students. The individual course success rates for QR and statistics went down, but they don't care about that. It's about the overall completion rates.

TABLE 7.1

Data for Fall 2017 at GCC with Standalone Developmental Math

Number of students who attempted introduction to algebra (who needed either QR or statistics)	Number (and percentage) of students who passed introduction to algebra (who needed either QR or statistics)	Number of students who attempted either QR or introduction to statistics with the booster course	Number (and percentage) of students who passed either QR or introduction to statistics with the booster course	Number (and percentage) of students who attempted introduction to algebra who eventually passed either QR or introduction to statistics with the booster course.
203	129 (63.5%)	432	260 (60.2%)	91/203 = 44.8%

TABLE 7.2

Data for Fall 2018 at GCC without Standalone Developmental Math

Number of students who attempted either QR or introduction to statistics with the booster course	Number (and percentage) of students who passed either QR or introduction to statistics with the corequisite	Number of students who would have placed into introduction to algebra (under the previous placement test scores)	Number (and percentage) of students who would have placed into introduction to algebra and passed either QR or introduction to statistics
652	335 (51.4%)	210	101 (48.1%)

Professor Smith provided his thoughts as to why the QR and introduction to statistics course success rates decreased, but the overall completion rates increased.

> You have some of the students who likely tested into the high-end of the introduction to algebra class, and that's why they likely passed the college-level courses, but I think we are doing such a disservice to so many students. There are students who really need that basic foundation [introduction to algebra], and by not offering that to them, we are setting them up for failure. I also think we lost some students who legitimately tested into QR or statistics [with the corequisite]. With teachers having to spend so much time helping underprepared students, maybe they didn't get the best instruction for what they needed.

In the fall of 2018, GCC removed all standalone remedial algebra and made college algebra with the booster course a default placement for students requiring college algebra. Tables 7.3 and 7.4 show the comparative results from 2016 to 2018.

Professor Smith clarified:

> It was such a disaster; we stopped it after that year. The problem was many of the students who would have tested into introduction to algebra or elementary algebra bombed college algebra. They were just completely

TABLE 7.3

Data for Fall 2016 at GCC with Standalone Developmental Math

Number of students who attempted introduction to algebra (who needed college algebra)	Number (and percentage) of students who passed introduction to algebra (who needed college algebra)	Number of students who attempted college algebra	Number (and percentage) of students who passed college algebra	Number (and percentage) of students who attempted introduction to algebra who passed college algebra
75	40 (53.3%)	229	116 (50.7%)	18/75= (24%)

TABLE 7.4

Data for Fall 2018 at GCC without Standalone Developmental Math

Number of students who attempted college algebra with the booster course	Number (and percentage) of students who passed college algebra with the booster course	Number of students of who would have placed into introduction to algebra (under the previous placement scores)	Number (and percentage) of students who would have placed into introduction to algebra (under the previous placement scores) and passed college algebra with the booster course
597	148 (24.8%)	82	5 (6.1%)

lost. In many cases, their grades of "F" were changed to a "U" [unsatisfactory] so that their grade point averages were not impacted. So, we still don't require the introduction to algebra course for the QR and statistics students, but we brought back the algebra sequence for the college algebra students.

Professor Smith added:

I just don't understand what people are thinking. Look at college algebra; there are polynomial functions, logarithms, and matrices. Who in their right mind would think that a student with hardly any math background could just learn the necessary skills and be able to apply them to the rigors of college algebra?

Arithmetic but No Algebra

In 2017, HCC decided to no longer require the 8-week long introduction to algebra course as prerequisite for QR and statistics with the booster course, but they kept the week-long arithmetic refresher. Moreover, students still needed a minimum arithmetic score to enter those courses. Was this structure sufficient?

No, it wasn't, *asserted Professor Bell*. Quantitative reasoning and statistics are college-level classes, and students need some basic skills and a certain degree of mathematical maturity.

Why is an arithmetic background not enough?

> Students need to be able to evaluate expressions. They need to be able to apply formulas. They need to able to solve basic equations. They need to be able to read and set up word problems, stated Professor Williamson.

Professor Hickey added:

> Yes, arithmetic is important, but they have to understand how to apply those concepts. They have to be able to read and interpret math and taking an arithmetic refresher or doing well enough on arithmetic is not enough.

Why the Push to Eliminate?

Developmental math was compressed; alternative and effective math pathways were established; students were moving through their math requirements at a quicker pace, and costs were cut. Why was there a continuing need to cut back further in developmental math? The faculty listed three common reasons.

> Our senior level administration is constantly worrying about losing enrollment to other schools, especially for-profit institutions. Our president is stunned how students pay thousands of dollars for for-profit schools when they could come to a community college for a much lower price. The answer is simple. For-profit schools let students skip a lot of general education classes and promise they will get jobs after graduation. So, they [students] take out loans thinking they will graduate quicker and get a job after they graduate. Students don't like developmental education classes, so less of them means more student retention, responded Professor Sutcliffe.

A second reason included mandates from the state:

> The elimination of standalone developmental math has been coming from the state, *asserted Professor Mussina*. People who don't know anything about math look at the data, and they see how students struggle in developmental math, so they assume developmental math is the problem.

Additionally, the faculty shared that overall dislike for developmental math pressured administrators into eliminating standalone developmental math.

> Our president and provost conducted these focus groups for students on their experiences at our school, *shared Professor Harnich*. Of course, the vast majority of them complained about taking developmental math.
>
> It even got to the point where students' parents were complaining about their children placing in developmental math. Yes, parents, exclaimed Professor Smith.
>
> Compressing and eventually eliminating developmental math became a reaction to political pressure. Students, and their parents, were the customers. They didn't want it [developmental math], so neither did the administration, stated Professor Holton.

By the mid-teens, Professor Thurmond had 35 years of experience teaching developmental math. She feels irrational decisions, such as eliminating standalone developmental math, stem from two major misconceptions regarding developmental education.

> First, people think that it's a mistake that students are placed into developmental math. Obviously, the ACCUPLACER isn't perfect, but by and large when students place into developmental math, they need to be there because their math skills are not college ready. Second, they think the reason students fail developmental math is our [the faculty's] fault. In other words, we're not teaching them well. That's why we have had so many top-down initiatives to increase student success. Administrators think if we just try the next initiative of the week, our students will succeed.

Slim Pickings for Non-Credentialed Faculty

How did the elimination of the standalone developmental math courses impact the non-credentialed faculty who were not able to teach college-level math courses?

> We are mostly teaching the business or allied health math classes for the associate degrees, *shared Professor Holton*. We are very limited as to what we can teach.

Some instructors are not happy with their assignments:

> We have to teach a lot of the booster classes for the QR, statistics or college algebra classes, *said Professor Sutcliffe*. I don't like it; I'm a professor, and I feel like I am working for the professor who teaches the college-level portion. Technically we're equals, but it feels like they are my boss.

Professor McDonald concurred:

> It's an insult [teaching the booster section]. The college-level instructors basically dictate to us what and how they want something taught. Then they get frustrated with us if it doesn't go well. I got a master's degree for this?

Habyan and Bordi Reverse the Course

In the fall of 2019, HCC brought back introduction to algebra as a prerequisite for QR and statistics with the booster course. In the spring of 2020, BCC did likewise and also brought back the 1-week pre-semester arithmetic refresher. Their faculty explained why they brought back standalone developmental math.

> It was the volume of student complaints, *explained Professor Hickey*. So many students in QR or statistics were going to the chair and the dean and complaining that they couldn't keep up with the course.
>
> As chair, I constantly had students in my office complaining these classes [QR and statistics] were too hard and the instructor was going too fast, *shared Professor Guzman*. But some quick research told me that those were the students who would have placed into algebra or arithmetic.
>
> We were getting parents calling into complain, *stated Professor Harnich*. They were upset that their children weren't being given a chance to brush up on their skills before college-level classes. I just wonder if those were the same parents complaining that their children were in developmental classes to being with.

There were even students who asserted that they needed standalone developmental math.

> It was some of the older students, *shared Professor Mussina*. They were like "Didn't anyone realize I needed some refresher math before taking these classes [QR and statistics]?"

The Pathways at the End of the Teens

Community colleges eliminated standalone arithmetic classes, designed alternative math pathways, and experimented with removing developmental math classes. The figures below show the math pathways for STEM and non-STEM students at BCC, GCC, HCC, TCC, and SCC at the start of 2020.

Griffin Community College

Non-STEM students at GCC can place directly into either the QR or introduction to statistics course with the corequisite without completing any standalone developmental math classes. Students who test high enough on the placement test do not need the corequisite component (Figures 7.4–7.5).

FIGURE 7.4
STEM students option 1.

FIGURE 7.5
STEM students option 2.

Bordi, Habyan, Telford, and Sisco Community Colleges

FIGURE 7.6
Non-STEM students.

FIGURE 7.7
STEM students option 1.

FIGURE 7.8
STEM students option 2.

Assisting ESL Students

In Chapter 4, the faculty discussed the large amount of ESL students, who struggled in community college math because of their language barriers. This became noticeable in the 1990s and continued throughout the aughts.

In fact, by the teens ESL students' struggles in developmental mathematics had become a national issue (Bailey et al., 2010; Boylan, 2011; Howard & Whitaker, 2011). Ultimately, there were two major reasons why ESL students struggled in community college math: First, for some students, their inability to understand English thwarted their ability to keep up with the class and learn math. Second, math is taught and expressed differently in other countries, and these nuances impeded students' learning of math in American community colleges.

The twenty-first century has witnessed a high volume of ESL students in American public community colleges, and researchers expect this trend to continue. For example, in 2011, a study in California reported that 25% of the students who attended public community college were ESL students (Llosa & Bunch, 2011). Additionally, a study in 2015 reported that English was not the first language for approximately 50% of the students who attended New York City's community colleges (City University of New York, 2016). Specifically, Hispanic and Asian students compose a large portion of the ESL students at American community colleges. In fact, nationally, 56% of all Hispanic undergraduate students attend public community colleges (Ma & Baum, 2016). In 2018, McFarland et al. projected a 26% increase in Hispanic students and a 12% increase in Asian students from 2015 to 2026.

A Deeper Understanding of ESL Students

Better serving ESL students started with better understanding them. Professor Bell elaborated:

> In the 1990s, and even into the 2000s, ESL students were just put in our classes and not much guidance was given to them or us. We did the best we could to teach them, and they had to swim or sink.

Professor Mesa explained how TCC took steps during the aughts and teens to better serve ESL students.

> It began with learning more about these students and how we can't simply group them into one category.

TCC offered professional development for faculty regarding ESL students. Professor Mesa explained how ESL students are categorized into two groups:

1. LEP (Limited English Proficiency): These are students who have very limited proficiency in English and who are unable to communicate in English.

2. Higher-lever ESL Students: For these students, English is their second language; however, they can communicate in and understand English. To some degree, however, they need to sharpen their English skills.

Professor Mesa discussed how these students are classified when entering TCC:

> When students enter Telford, they are given a written test. Also, ESL students are invited to meet with an ESL specialist upon admission. This person [the specialist] has a master's in ESL. Through both assessments, we are able to determine the students who need the most help [LEP students] and students who have some proficiency in English and to what degree of proficiency that is.

Professor Mesa elaborated as to how this helps place students:

> The LEP students need to complete a basic skills class in English before taking a math class. This class is for students who have very limited skills in English. The class focuses on basic reading and writing but also basic English conversation. I guess the class is designed with the goal of "What are the basic English skills that a student needs to survive in a college class?"

Professor Mesa continued:

> For the ESL students who are more proficient in English, we have elementary and intermediate ESL classes. These are for ESL students who have basic conversational and writing skills in English but need help sharpening those skills.

Additional Resources for ESL Students

Like TCC, HCC has also tried to improve the educational experience for ESL students. Also, like TCC, HCC classifies incoming ESL students and assists those with limited English proficiency. Professor Williamson elaborated:

> With the help of a grant, our college invested in helping our ESL students. Part of it was hiring more advisors and academic coaches with backgrounds in working with ESL students. We also developed a peer mentoring program. Students at Habyan, for whom English was a second language, are matched with incoming ESL students. These mentors not only help these students become more fluent in English; they help them adapt to community college life and help them to feel more comfortable.

Professor Hickey discussed the outcome of this program:

> We've had some very positive results. ESL students have reported feeling more comfortable. Working with advisors and mentors really helps develop better conversational skills. It helps them understand their instructors better and communicate with their classmates better.

Professor Bell discussed how this type of advising and mentoring is imperative for students entering community college math:

> Students need basic conversational skills. They need to be able to understand their instructor. They need to be able to communicate with their classmates. The bottom line is we have to find ways to help them become proficient, or proficient enough, in English to be successful in math.

Professor Williamson added:

> They need to develop conversational skills before they take a math class and continue developing them while taking a math class.

Professional Development for Math Faculty

To assist HCC faculty better serve ESL students, the college has offered professional development for faculty. Professor Hickey clarified:

> The workshops have been great. We have ESL students tell us about their experiences and what it's like starting at a community college when English is not your first language. We get to hear about some of the anxieties they face and how learning math in a different country can compound their math anxiety.

Professor Williamson discussed how this has helped educators:

> I think it's made us more sensitive to ESL students. We go that extra mile to ensure that they are acclimating to our math classes and if they're understanding everything or need anything.

Professor Bell added:

> Oftentimes, these [ESL] students are shy or nervous about reaching out for help, so us reaching out to them really helps.

Professor Williamson discussed how these workshops have helped him:

> In some of the workshops, we have learned about how students learn math in other countries and how it's different from how we teach math. It's been really eye-opening for me. It's just given me a better understanding of where these [ESL] students are coming from.

Professor Williamson shared how this has changed his assessment style:

> I hate to admit it, but especially in my developmental math classes, I used to insist that they do the problems a certain way. I did this for their own good. I just wanted them to develop a certain structure,

but I realized that I wasn't always being fair. Students, especially ESL students, come into our classes having learned math different ways, and we need to be flexible.

Professor Hickey concurred:

Students thought I was evil because I would require them, in my developmental math classes, to do problems my way and show all their steps. My mindset was simple. I figured they were in these [developmental math] classes because they had bad work habits and bad organizational skills, so I was going to correct them and get them on the right track. I still think, as teachers, we need to help students develop good math organizational skills when doing problems, but I do realize we need to be flexible. We can't assume that everything they have done before our class is wrong. We need to respect their backgrounds and work with them.

The faculty discussed how they have altered their teaching style to help ESL students.

I became more careful to not to talk as fast and to annunciate more. Instead of just asking questions verbally, I would show them [the questions] on the board or the screen sometimes. It's just a way of being more sensitive to students' learning needs, stated Professor Noles.

In the past, I would ask too many open-ended questions, and we learned that ESL students struggle with and can become intimidated by open-ended questions, *stated Professor Hickey*. So, I altered my questions.

Professor Hickey provided some examples comparing open ended questions with specific questions:

Open Ended Question: When computing long division of polynomials, ask "How do we simplify this?"

More Specific Question: Ask various students, "What is the first thing you do? What is the second thing you do?" And so forth.

Open Ended Question: When computing a probability question with "and" or "or," ask "What formula do we use?"

More Specific Question: "Do you see any key words?", "Are these events mutually exclusive?" "What does that mean?" "From all this, do you think you know the formula to use?"

Professor Hickey added:

In hindsight, using these kinds of questions and giving this kind of assistance is something we should be doing with all our students. Learning about ESL students has just made many of us more sensitive to students in general.

Continued Challenge

While community colleges have made progress serving ESL students, challenges remains:

> There is such a high volume of ESL students each year, *shared Professor Mesa.* It's just so hard to ensure that all their needs are met.
>
> We're doing everything we can to assess incoming ESL students, place them properly, and set them up with advisors and mentors, but how do we help accommodate the large and growing number of ESL students? I don't know the answer, reflected Professor Noles.

Addressing the Equity Gap for Minority Students in Math

As discussed in Chapter 5, the overrepresentation of Black and Hispanic students in developmental math surfaced in the 2000s. Community colleges and math departments faced pressure to remedy this situation, but faculty and administrators did not know how. Fortunately, with further research and collaboration, schools began addressing these equity gaps during the teens.

Where Are the Minority Students Coming From?

The faculty and administration at TCC realized that to better serve minority students, they needed to understand them better. The TCC faculty as well as the SCC faculty had the same realizations regarding Black and Hispanic students. Professor DeSilva elaborated:

> We spent too many years wringing our hands over too many African American and Hispanic students in developmental math and not getting out of developmental math. It was time to stop talking and start doing.

Professor DeSilva clarified:

> There was a college-wide initiative to invite African American and Hispanic students to focus groups. Honestly, the goal was just to try and understand them and their needs better.

Professor Moyer, a strong advocate for minority students, shared the findings of these sessions:

> Several students mentioned that they came from poor neighborhoods and inner-city schools. They discussed how the problems at their schools were so big, like behavioral problems, drugs, and even gangs. Learning just wasn't a priority.

> I don't know if anyone ever computed the percentages or did the math on how many, but we found many of our Black and Hispanic students, who placed into DEV math came from the inner-city. They wanted more from life and with low tuition and [financial] aide available, community college was their best option, asserted Professor Mesa.

Professor Noles, who previously taught in an inner-city high school, confirmed this:

> In many inner-city schools, the priorities are just different. It's about survival. Learning is secondary. So many students actually come to school and are just scared for their lives. They're afraid to even walk the halls because they might get beat up. Classes are overcrowded so students don't always get the attention they need.

Professor Noles added how the homelife of a minority student impacts their education.

> Many of these [minority] students come from single-parent households with no father. They have to work to help their mother's take care of their brothers and sisters. To be honest, it was always a hard concept for me to grasp. When I was in high school, my parents took care of everything. All I had to do was be a kid and go to school and learn.

Professor Trombley, another former inner-city schoolteacher, explained how these hardships impact learning math:

> If it's been said once, it's been said a million times, math is linear; math is progressive; math is unforgiving. If you fall behind in math and develop gaps, it's so hard to catch up. I worked with so many high school students who had gaps in arithmetic and basic algebra, and we just did the best we could to get them through high school. When students face the kinds of problems that exist in the inner-city, you don't learn math consistently, and that inconsistency hurts them in the long run.

It is noteworthy that on a national level, Black and Hispanic students have the lowest success rate, among all ethnic groups, in mathematics from kindergarten through 12th grade (Bahr, 2010). Consequently, many of these students are behind academically when entering community college.

Outreach to the Inner-City Communities

Many of the problems that minority students face in community college math begin long before they enter college; therefore, both SCC and TCC designed a student outreach program for minority students at local inner-city schools.

The program was aimed at minority students in their junior or senior year who wanted to attend college but needed academic assistance.

The first step is to identify the students who want to participate, and we let the schools take care of that, *reported Professor Mesa.* Next, these students are given a math assessment. It isn't a computer-adaptive test. It's a test they take by hand that we [the math department] created. It's divided into sections, you know arithmetic, basic algebra, intermediate algebra. It's graded by hand to see where they need help and where their gaps are.

Who grades the exams?

We have peer tutors who do that. These are math tutors who work in our tutorial center.

Professor DeSilva explained how the program operates:

Each week we have some math faculty and some of our math tutors go into the schools and work with these students after school. We have all kinds of worksheets, and they even use ALEKS [math software] as well. We also direct them to online math videos as well. The goal is to get them to test out of developmental math when they come to our school.

Professor Moyer discussed how this program has positively impacted the students:

You can tell so many of these students are just appreciative that we care about them and that we are going out of our way to their school to help them.

However, it is not easy.

I always knew these [minority] students had major gaps, but I never knew how much. The students are lacking in algebra but also arithmetic and some even in basic arithmetic skills like multiplication and division. For many of those students, it's not enough.

Has this outreach program been successful? Are less minority students placing into developmental math?

We've seen a small improvement. We still have way too many [minority] students placing into our DEV math classes, but it has been a start. We have to find a way to help more students before they come to community college, reported Professor Noles.

Professor Trombley concurred:

We've closed the gap a little. I think in many cases [minority] students are still placing into developmental math but higher up, so that is something. The need is just so great. It's going to take time. The bottom line is K-12 schools need to find a way to provide earlier intervention to students falling behind in math.

Assistance and Guidance When Entering School

TCC, SCC, and HCC offer mentoring programs for Black and Hispanic students.

> It's like our work with the ESL students, *stated Professor Bell*. Incoming Black and Hispanic students, who are interested, are paired with a student or faculty mentor.
>
> It's someone who meets with them, keeps in contact with them and helps them acclimate, *reported Professor DeSilva*. We're trying to ensure these students feel comfortable and don't fall through the cracks.
>
> I'm a mentor for African American students and you bet I keep on my people about how they are doing in math, and if I get the feeling they are struggling, I get on them, about getting help right away, asserted Professor Moyer.

Educating Faculty

HCC has taken steps to educate their faculty on ESL students, and HCC along with TCC do the same for their faculty regarding Black and Hispanic students as well.

> I have to admit; I was resistant at first, *admitted Professor Hickey*. I was like; it's math. What's the big deal? If we teach them well, and they want to learn, they will pass, but I kept an open mind, and I realized it's not that simple for African American and Hispanic students.

Professor Hickey elaborated:

> Many of our students have had bad experiences in math, but this has been especially the case for minority students. Many of them feel they will fail math even before starting.

Professor Douglass discussed a revelation:

> I never realized that African American students, particularly, were less likely to ask questions when they need help. They haven't really been in situations where teachers have reached out to them before, so they think they are on their own. It's made me more sensitive to their needs.

Professor Trombley shared how trying to understand her students has reshaped her teaching:

> I'm more sensitive to my students' background an interest. At the beginning of the semester, I give my students a questionnaire where I ask them to tell me a little bit about themselves. What are their interests in sports, movies, TV shows and whatnot? I do this because I learned how minority students can feel marginalized in a math class, and I want them to feel included.

Should minority students be taught math differently than White students? Professor DeSilva had a strong response:

> As a Black woman, this question angers me. In fact, I overheard a conversation between two students after class one day that annoyed me. A Black student was complaining about struggling in class, and to be honest, this guy was struggling because he wasn't doing what he was supposed to do, you know doing homework or coming to class enough. The White girl said, "I can tutor you because I have a very Afrocentric way of explaining things." I'm thinking, "Are you kidding me?" As a Black person, I was insulted. I have a bachelor's and a master's degree in math. I had White teachers, Black teachers, and Asian teachers. I succeeded in math because I came into the class with the proper prerequisite skills; I worked hard and got help when I needed it, and I had good teachers. The end.

Professor Moyer summarized:

> Here's the deal. Students need to master the same math concepts. We can't do certain students more favors than others, but we need to understand more about our students. We need to understand where they are coming from and more about their backgrounds and their needs. The better we understand our students, the better we can teach them.

General Student Outreach

For decades, community colleges have been reaching out to students to help them outside of the classroom. Student outreach has continued to develop over time, and by the late teens, community colleges had implemented programs to assist students before they even register for their math classes.

Student Profiling

By the teens, students had a variety of modalities to choose from when registering for math class, and they also had a variety of math pathways open to them. While such flexibility is beneficial to a heterogeneous group of students, the variety can be overwhelming.

Both TCC and SCC utilize a form of student profiling to assist advisors in helping students navigate their pathways. At both TCC and SCC, students must meet with their advisors before registering for classes in their first semester. The math faculty at each school have designed a questionnaire for advisors to administer to students.

This questionnaire allows the advisors to help the students choose their pathway and modality. For example, students who prefer traditional instruction are steered away from the emporium model. There are faculty from both math departments who serve as advising liaisons. College advisors contact these faculty members when more information is needed, for students, to select a math pathway. Professor Douglass elaborated.

> We get students who test into introduction to algebra, which means they have weak math skills, but they want to pursue these advanced fields in science like forensics or physics. Some [students] even want to go into engineering. That's great, but in many cases they're not aware of how much math is involved. So, it's not that we want to drive them away. We want to make them aware of how much math they will need and the long [math] pathway involved. In some cases, the student switches to either the QR or statistics pathway.

If students choose the QR or statistics pathways, the advisors and faculty liaisons work with them to determine which is best. Professor Trombley explained:

> We want to make sure students take a course that aligns with their career pathway and will transfer to a four-year school. For the students who express interest in the behavioral or social sciences, we direct them toward statistics. For the students who are looking toward the fine arts or majors like English or sign linguistics, we direct them toward quantitative reasoning.

The advisors at TCC make an effort to get to know math faculty members' teaching styles and consequently recommend certain instructors for certain students. Professor DeSilva provides an example:

> Some like to do group-work more than others, so advisors will ask students about that and recommend accordingly.

At SCC and TCC advisors and the math faculty work especially hard to direct the arithmetic students. Professor Moyer explained:

> We really try to be careful with the students who test below algebra. We know they aren't ready, but we don't want to lose them. For the students who don't miss [score on the placement test] the introduction to algebra class by much, we recommend they take the arithmetic refresher class. If they want more time and a slower place, the advisor sets them up in the tutorial center for the individualized program. For the low-level students, they are directed to the adult basic education class.

Professor Douglass added:

> When students enroll in the adult basic education class, our advisors keep in contact with them. They [the advisors] will call them every so often to see how they are progressing.

Early Intervention

Since the 1980s, community colleges have sought to help students who are struggling in their math classes. This attention continued in the teens, with an emphasis on early intervention.

We have an early intervention program at Griffin [Community College], *shared Professor Timlin*. After two weeks, we submit a list of names of students who are struggling electronically, and the information goes to the right advisor or coach or intervention specialist. That person then reaches out to them to see how they can help them.

The key is to get to the students as soon as possible, *stated Professor McDonald*. It's very important to get off to a good start in a math class.

Internet Videos

With the establishments of YouTube and Khan Academy, which was spear-headed by Sal Khan, math videos on the Internet became popular in the 2000s.

It seemed like early in the 2010s [teens], we were getting more and more students who were relying on Internet videos to help them with their homework, *shared Professor DeSilva*. I guess it was something they started doing in high school.

Yeah, around 2012 or 2013 I think it was, I got more and more students asking me about what videos they could watch online to help them. Honestly, I had no idea. I hated that question, because I was so unfamiliar with Internet videos, shared Professor McDonald.

What makes a quality video for a student seeking help in community college math?

It needs to be straight to the point, yet thorough. For example, if it's a video on using the quadratic formula, remind the student what constitutes a quadratic equation; show the formula; walk the student through substituting the numbers for the variables; show the simplification, and give a brief description for the two solutions, responded Professor Guzman.

The tricky thing is the videos can't be too long or cumbersome, because this generation [later Generation Y and Generation Z] loses interest very quickly. So, it has to be quick but complete and not use a lot of fancy terminology. Just guide the student through the problem step by step, shared Professor Sutcliffe.

Professor Noles added:

It's important to remember that YouTube videos or Khan Academy videos aren't meant to teach students from scratch. They are meant to help students who are struggling but have some background information. They could be review problems as well, but you can't rely on them [the videos] as the sole source of instruction.

Professor Mussina asserted that community college math instructors have a role and a responsibility regarding Internet videos.

> Unfortunately, there is a lot of garbage on the Internet. Anyone can make a video and post it online. Heck, the math might not even be correct. So, as faculty, we have the responsibility to do our homework and let students know which videos are good quality. I mean, it can be subjective. What I like, another professor may not like, but I think we can agree on what is good quality. So, it's on us to do our research and inform our students.

Open Educational Resources

Throughout the twenty-first century, cost has been a recurring issue in higher education. However, this was not limited to reducing college costs. There has been an effort to reduce costs for students as well.

Reduced Cost or Free Software

As enrollment in community college became competitive in the twenty-first century, administrators became concerned with the cost of college textbooks and how this impacted students. Software programs under companies such as Pearson and McGraw Hill came under fire for their high costs. For example, students may have had to pay more than 100 dollars for a My Lab Math access code for a particular math course.

> Our dean called a meeting with a couple of us in 2015, and she started talking about open educational resources and how we had to make things cheaper for students, *recalled Professor Harnich.* We were immediately frustrated. We had just gotten everyone good with MyMath Lab, and she wanted more change.
>
> I just figured it was another case of student consumerism, and our administration's efforts to compete with other schools by lowering costs, shared Professor DeSilva.
>
> Our immediate concern was that our administration was putting cost over quality. We knew programs like MyMath Lab and ALEKS were expensive, but we liked what they had to offer. We liked the interactiveness with students; we liked the embedded videos; we liked the tech support from the publishers, asserted Professor Hickey.

As the teens progressed, faculty began researching and experimenting with math open educational resources. This created some frustration at first.

The quality just wasn't there. We looked at some programs, and they were just too clunky, *shared Professor Mesa*. We didn't think the students were getting enough feedback, and the tech support was basically non-existent.

However, BCC, TCC, and HCC discovered an open educational resource that has met their students' needs, My Open Math. Professor Moyer elaborated on why this is:

MOM [My Open Math] is just great. The students get the instant feedback, which is the most important thing. The individual videos for each question are amazing and clear. There is still a really good online gradebook. It's like they are getting everything they did with My Lab Math [formerly MyMath Lab], but MOM is free.

What about the recurring issue of tech support?

There is no professional tech support, *shared Professor Guzman*. There is an online community on the My Open Math site where instructors help each other out. I've asked many questions about the logistics, and I get clear and concise answers.

Professor Williamson relayed an advantage that some open educational resource programs have over programs with traditional publishers.

With programs like My Lab Math, you are tied down to one book. With OERs [Open Educational Resources], you can select questions from several textbooks. When I designed our QR course with My Open Math, I could pick from several QR books and even some statistics books.

I asked the faculty if they think that open educational or reduced cost resources will become the norm for community college math:

The My Open Math software has worked well. Knewton has worked well for us as well, but I'm not 100% sold. The actual textbooks we have used [open educational resources] have not been as good. There have been mistakes in the sample problems and the overall quality isn't as good [as commercial products], shared Professor Guzman.

With companies like Pearson, you get textbooks that are more catered to your courses. We used an OER [open educational resource] book for our college algebra class, and there was just too much content in the book. There were so many topics we didn't cover. It was confusing to the students, reflected Professor Hickey.

You get what you pay for. Yes, traditional textbooks are expensive. But you are paying for a quality product that meets the needs of your students, shared Professor Moyer.

Professor Harnich summarized:

I get it; textbook costs are high. I remember when I was a student, but we need software that is high quality and meets the needs of our students. Students need good questions, and they need to know right away if they

got a question right or wrong. They need good videos. They also need
an interface that they can easily navigate. They need technical questions
answered correctly and right away. It's that simple.

As the teens concluded, the effectiveness of open educational resources in
community college math was inconclusive.

Summary

Reform for community college math continued during the teens. Standalone
developmental math courses were compressed and even eliminated. Instead
of requiring students to work up to and complete college algebra, many
schools adopted alternative math pathways where students could do less or
no remedial coursework and then attempt QR or introduction to statistics.
Corequisite education exploded, as students were able to complete the remedial component in tandem with the college-level portion.

Alternative math pathways have positively impacted students overall.
Students no longer need to take long sequences of developmental math only
to attempt college algebra, a course that is a poor fit for non-STEM majors.
Students are moving through their math requirements at a quicker pace.
QR and introduction to statistics also allow students to contextualize math.
Unfortunately, the changes to math pathways have gone too far. Some stu-
dents still need standalone developmental math to provide them with the
foundation necessary for attempting a college-level course with a corequisite.

The teens were a time of continued change and reform. Some of this change
positively impacted community college math and could become a permanent
fixture in the discipline. Other initiatives will likely fade. Community college faculty continued to contend with an ever-evolving discipline. However,
nothing could prepare the faculty, or anyone else, for what loomed at the
start of the 2020s.

References

Altose, A (2018). Embracing the value of college math. *Inside Higher ED*. https://www.
    insidehighered.com/views/2018/10/11/why-higher-ed-needs-new-approaches-
    teaching-math-opinion.
Bahr, P. R. (2010). Preparing the underprepared: An analysis of racial disparities in post-
    secondary mathematics remediation. *Journal of Higher Education, 81*(2), 209–237.
Bailey, T., Jeong, D. W., & Cho, S.-W. (2010). Referral, enrollment, and completion
    in developmental education sequences in community colleges. *Economics of
    Education Review, 29*(2), 255–270.

Boylan, (2011). Improving success in developmental mathematics: An interview with Paul Nolting. *Journal of Developmental Education, 34*(3), 12–41. https://files.eric.ed.gov/fulltext/EJ986275.pdf.

Byrk, A. S., Gomez, L. M., Grunow, A., LeMahieu, P. G. (2015). *Learning to improve: How America's schools can get better at getting better.* Cambridge: Harvard Education Press.

Cafarella, B. (2021). *Breaking Barriers: Student Success in Community College Mathematics.* CRC Press.

Carnegie. (2020). *Quantway.* https://carnegiemathpathways.org/quantway/.

City University of New York. (2016). *A profile of undergraduates at CUNY senior and community colleges: Fall 2015.* New York: CUNY Office of Institutional Research and Assessment. http://www2.cuny.edu/wpcontent/upload/sites/4/mediaassets/ug_student_profile_f15.pdf.

Clyburn, G. (2013). *Improving on the American dream: Math pathways to student success.* https://www.carnegiefoundation.org/wp-content/uploads/2013/09/Improving_on_the_American_Dream.pdf.

Complete College America (2021). No room for doubt: *Moving corequisite support from idea to imperative.* https://completecollege.org/wp-content/uploads/2021/04/CCA_NoRoomForDoubt_CorequisiteSupport.pdf.

Elrod, S. (2014) Quantitative reasoning: The next 'across the curriculum' movement, *Peer Review, 16*(3), 4–8. https://www.aacu.org/peerreview/2014/summer/elrod.

Howard, L, & Whitaker, M. (2011). Unsuccessful and successful mathematics learning: Developmental students' perspectives. *Journal of Developmental Education 35*, 2–16.

Lewy, E. B. (2021, March 12). *Barriers to scaling up corequisite classes and multiple measures assessment.* Center for the Analysis of Postsecondary Readiness. New York. https://postsecondaryreadiness.org/corequiste-multiple-measures-scaling-barriers/.

Llosa, L., & Bunch, G. (2011). *What's in a test? ESL and English placement tests in California's community colleges and implications for us-educated language minority students.* Menlo Park, CA: William and Flora Hewlett Foundation. https://escholarship.org/uc/item/10g691c.

Ma, J., & Baum, S. (2016). *Trends in community colleges: Enrollment, prices, student debt, and completion (College Board research brief).* New York: The College Board.

McFarland, J., Hussar, B., Wang, X., Zhang, J., Wang, K., Rathbun, A., Barmer, A., Forrest Cataldi, E., & Bullock Mann, F. (2018). The condition of education 2018 (NCES 2018–144). Washington, DC: U.S. Department of Education, National Center for Education Statistics. https://nces.ed.gov/pubsearch/pubsinfo. asp?pubid=2018144.

Merseth, K. M. (2011). Update: Report on innovations in developmental mathematics – Moving mathematical graveyards. *Journal of Developmental Education, 34*(3), 32–39.

Park-Gaghan, T. J., Mokher, C. G., Hu, X., Spencer, H., & Hu, S. (2020). What happened following comprehensive developmental education reform in the sunshine state? The impact of Florida's developmental education reform on introductory college-level course completion. *Educational Researcher, 49*(9), 656–666. doi: 10.3102/0013189X20933876.

Smith, A. D. (2019). Relationship between required corequisite learning and success in college algebra. *Georgia Journal of College Student Affairs, 35*(1), 23–43. doi: 10.20429/gcpa.2019.350103.

Vandal, B. (2016). Coreq and college algebra. *Complete College America.* https://completecollege.org/article/coreq-and-college-algebra/.

8

Teaching during the Pandemic: What We Experienced and What We Learned

Monday, March 9, 2020. While it was an ordinary day, I will never forget it. I had just returned from spring break and was gearing up for the final 8 weeks of the semester. I taught two classes that day, and at the end of each class, I remember telling the students that I would see them on Wednesday. I never thought that would be the last time I would see my students that semester. The next day, March 10, our school would be shut down due to the Coronavirus pandemic. I never thought those would be the last face-to-face classes that I would teach for over 2 years. Moreover, in the coming months, I never thought so many hundreds of thousands of lives would be lost.

A deadly virus spread across the globe like wildfire. While people had to isolate and engage in social distancing to save lives, community colleges still needed to meet students' educational needs. Students came to us for an education and providing them with every opportunity to achieve that education was our responsibility. However, we had a problem. How could we best serve our students if we could not teach them in a face-to-face manner?

Community College Math Goes Fully Virtual and Online

Distance learning, which later evolved into online learning, has grown steadily since the 1990s. In fact, by 2018, nearly one-third of higher education students were enrolled in at least one online course (Online Learning Consortium, 2018). Therefore, online learning was not new to higher education when the Coronavirus shutdown occurred. However, virtual learning was quite new to higher education, especially community college math, and schools had no choice but to adapt and to do so quickly.

Different community colleges may have slightly varying definitions regarding virtual learning. Virtual learning is generally when students can access a course via the Internet; however, virtual learning differs from online learning in that in virtual learning students learn synchronously as a group. More specifically, in virtual instruction, students participate in live and interactive instruction with their classmates. Like face-to-face instruction, classes cover specific content on specific days, and students even take their exams at the same time. Conversely, online learning consists of students learning asynchronously and at times at their own pace.

The Participants

In this chapter, Professors DeSilva (Sisco Community College), Guzman (Bordi Community College), Holton (Griffin Community College), Hickey (Habyan Community College), Williamson (Habyan Community College), Noles (Telford Community College), Mussina (Bordi Community College), Moyer (Telford Community College), Sutcliffe (Griffin Community College), Mesa (Telford Community College), McDonald (Griffin Community College), Smith (Griffin Community College), and Harnich (Bordi Community College) shared their experiences. Additionally, I presented the experiences of three new faculty members.

Professor Brown

Professor Brown began teaching, full-time, at Bordi Community College in 2018. She teaches both developmental and college-level math and has a bachelor's and a master's degree in mathematics.

Professor Rhodes

Professor Rhodes started teaching at Telford Community College in 2019. He has a bachelor's and a master's degree in math and teaches both developmental and college-level math.

Professor Stephens

Professor Stephens started teaching at Telford Community College in 2019. She has a bachelor's and a master's degree in math.

Conversion to Virtual Instruction

When the pandemic hit, community colleges had to find a way to help students in face-to-face classes.

> Students did not sign up for an online course, and a lot of our students have a great deal of math anxiety anyway, so we had to find a way to give them live instruction with guided practice, because that is what they need, asserted Professor Brown.
>
> Our administration wanted us to use virtual instruction, where we were working with students synchronously. They didn't want us to just post videos or assignments and then have students contact us if they needed to, and we agreed, stated Professor Rhodes.

Most of the community colleges, in this study, utilized Zoom: Video Conferencing for working with their students. Bordi Community College started employing Google Hangouts but later switched to Zoom.

> Zoom really became the national way to communicate with people during the pandemic. You turn on the TV and all the interviews were being conducted by Zoom, stated Professor Guzman.

Coming to the decision to convert their math classes to virtual instruction using Zoom was the easy part. The difficult part was getting the rest of the faculty familiar with Zoom in a short time.

> We had to train our entire department on how to use Zoom, all the features of Zoom, and using a document camera so you could show the problems. It was hard to believe how many faculty weren't familiar with Zoom, shared Professor DeSilva.

Professor Noles concurred:

> It was surprising. I had been using Zoom to communicate with students during office hours when they couldn't come to campus. We would talk on Zoom if they had a question on a homework assignment, and I couldn't explain math by email. I thought more people were doing it.
>
> We had good teachers who weren't up to speed with technology, stated Professor McDonald.

This lack of technological knowledge went beyond Zoom.

> Some of our faculty were struggling with basic things, like how to attach their notes or PowerPoint presentations to a My Lab Math or a My Open Math interface, shared Professor Guzman.

Professor Mesa added:

> After we showed our instructors how to record a video using Zoom, some really struggled with how to save it as a link and upload to the My Open Math interface.

Of course, this was a learning experience for even the faculty who were more skilled in technology:

> I'm embarrassed to admit it, but I was not familiar with Desmos [online graphing tool]. I had been teaching college algebra, and I was still graphing polynomial functions by hand, but another professor did a workshop on different ways we could teach students, and I couldn't believe what I was missing, stated Professor Rhodes.
>
> A colleague of mine showed us Screencast-O-Matic. You know when you need to show your students how to navigate something, like how to create a scatter plot using Excel. Screencast is great for that, shared Professor Smith.

Lack of Engagement

Throughout the history of community college math, faculty have stressed the importance of student engagement. There is not one correct way to engage students; there are several, and faculty must find what works. However, the faculty commented that engaging students in a remote environment is challenging.

> Most of the students didn't keep their web cams on during class. We required it for the exam, which was in advance, but there was ambiguity as to whether we could require them on a daily basis. I could ask the students to turn them on [the web cameras], but they didn't always do that. It's harder to engage students when you can't see them, said Professor Rhodes.
>
> I never realized how much I rely on my students' expressions and body language to guide how I teach or engage them, *shared Professor Holton*. If I see that students are not getting a topic [in a face-to-face class], I may alter my teaching approach. Sometimes, I have stopped class and ask students to write down on a sheet of paper what they are struggling with, so I can better help them. When I see students starting to drift or becoming fidgety, I know it's time to change things. I might start a problem and call on different people to answer certain parts of it. I might assign a problem and put them into groups. But again, not being physically there with the students, I lose that ability.

Being in-person also allows instructors to read positive cues as well. Professor Williamson commented:

> When you are in the class [in-person], you can see if students are getting it. You can read facial expressions and how students react when making eye contact. You can walk around as you give them problems and look at their answers. The bottom line is I know when it's time to move on to another topic. With the virtual classes, sometimes, I think

I am spending too long on a topic or doing too many sample problems, because I can't see if they are getting it or not. Then, by spending too long on a topic, I am losing them.

Professor Brown concurred:

I would call on students; sometimes they would answer; sometimes they wouldn't. I would try putting students in the Zoom break out sessions, but it just wasn't the same as being in class. I felt like it was all out of synch. Not being there in person, I lose the ability to engage the students effectively.

However, in some cases, the students were behaving in ways that made engagement impossible.

The problem is with the remote classes, students do too much multitasking. I figured out that students were at their jobs and just having our class on but not really paying attention. That's not good for any class but especially not college algebra, stated Professor Harnich.

Professor Brown agreed:

They would log into class and just do other things. I even had students admit they would log into class and do homework for another class. Learning math requires concentration. Yes, they have to practice after class, but they need that initial guidance from the teacher.

As Professor Brown mentioned, Zoom allows the use of break rooms. This is where the moderator can assign students into smaller groups to work on an activity. The moderator can join each group to supervise or answer questions.

It's a great feature, and I'm glad we have the option to break students into groups, but it's not always the same as in-person classes. Students are much less likely to talk to each other when I put them into groups. We have those students who sign into class, then walk away, so they don't even know they are in a group, opined Professor Brown.

Professor Rhodes agreed:

Many students don't want to turn on their web cams unless they absolutely have to, so I think not seeing each other face-to-face decreases the motivation to participate. Not turning on the web cams makes it less likely they will share their work with each other, which defeats the purpose of cooperative learning.

Professor Williamson added:

So many times, I assign the breakout sessions, and I will get one student who requests to be switched to another group because no one else is talking. But I guess that's better than when I go into a breakout room, and no one is talking.

The participants asserted that whether it is virtual or online instruction, it is the responsibility of the faculty to try and engage the students and help the students feel included in the class. Professor Stephens elaborated:

> Students can feel a sense of isolation in online classes or even virtual classes. If they don't feel included, they can lose motivation or stop trying. Faculty need to reach out to students individually by email to check in on them. Faculty should also set up discussion groups, so that students can communicate with each other.

Volume of Need Too High

While some instructors had difficulty keeping their students engaged, others were overwhelmed by the volume of questions they received during a class. This was the case for instructors who facilitated developmental math classes in the emporium model via Zoom.

> Normally, we have one instructor for about 50 students in a class, but we also have two in-class tutors to help us. Problem was when COVID hit, the tutors were laid off to cut costs. So, I was on my own with 50 students, *shared Professor Harnich*. Since the class was self-paced, I would open up a Zoom session, and students could drop in at any time with questions.

Professor Mussina discussed the complications of this situation:

> Me and the other [emporium model] instructors were just overwhelmed. We were getting so many students with questions. What we would do is have them sign in, we would take notice and assign them a number, like calling customer service, and we would help them in that order.

Professor Mussina added:

> The emporium model is just not a fit for Zoom. To be an effective instructor in the lab [emporium model], you walk around the room and watch the students as they complete the assignments. You talk with them and ask them questions and even catch mistakes as they are making them. You can't do any of that on Zoom. I mean, students can share their screens with you, but generally you don't know what they're doing.

Professor Guzman discussed how the emporium model via Zoom defeated its purpose:

> We have a computer lab that is basically open all day. Students can come in anytime, even if it's not their class and get help. That was the main purpose of the emporium model. But you can't do that with Zoom. One instructor can't let hundreds of students into his or her class, so students really only get help during their specific class time.

How did the Bordi Community College faculty accommodate students?

> Many of us just set up extra time during the day, other than office hours where we could help students, *responded Professor Mussina*. I would have students text me if they had a question, and if I was available, I would jump on Zoom. It was basically like I was on call all the time.

Did virtual instruction for the emporium model impact success rates?

> Our attrition rates went up. More students were dropping our classes because they couldn't get the help they needed, *answered Professor Guzman*. Never again. No way are we ever doing the emporium model like this again.

Griffin Community College (GCC) had been employing a hybrid model where part of developmental classes was lecture and the other part was in a computer lab. How did GCC accommodate this?

> We knew right away that the lab model wouldn't work on Zoom, so during the pandemic, we lessened the questions they would have on My Lab Math and did more explanation during the Zoom sessions, *responded Professor McDonald*. You can't do a self-paced lab on Zoom.

Testing

Mathematics typically utilizes proctored in-person testing to ensure validity. Even online students generally need to report to some type of testing site to take their exams. Some schools even employ remote proctored testing programs such as Respondus and ProctorU. However, given the short notice of the pandemic shutdown, schools did not have time to come up with a testing plan. During the spring semester of 2020, Sisco, Bordi, Telford, Habyan, and Griffin Community Colleges decided to simply give students un-proctored take-home exams.

> It wasn't ideal, but it was the best we could up come up with, *stated Professor Mussina*. We had it set up where we could release the test on our course's web page; students could access it, complete it and scan it back to us within a certain amount of time. Obviously, we couldn't watch them take the test or even know if it was them taking the test.
>
> Were we afraid of cheating? Of course, we were, but we were trying to make things as manageable [as possible] for the students, *asserted Professor Moyer*. What some of us did, knowing students would be using their notes, was give harder exams and require that students show their work.

This was a challenge for the developmental classes:

> In our algebra classes, you really can't make the tests harder, you have to just test them on the content that is covered. We were worried about cheating because we need students to be prepared for the next class, stated Professor Sutcliffe.

TCC faculty attempted to have their tests proctored where students would take their exams on Zoom with the web camera focused on them taking the exam so that their instructors could watch.

> It sounded like a great idea, but there was one problem. So many of our students didn't have a functioning web cam[era] or a web cam at all. So, that was the end of that. Again, they didn't sign up for this, recalled Professor Mesa.

Did the faculty suspect an increase in cheating in this time?

> My pass rates increased a little but not much. Most of the students who had been failing my class continued to do so, shared Professor Noles.
> I honestly didn't notice too much of a difference. They could use their notes, but with math you have to know how to do each type of problem. For example, if the question is on factoring trinomials and you look at your notes and see another example of factoring trinomials, but that doesn't mean you can do the problem I gave you, responded Professor Sutcliffe.

Professor Brown had a specific concern regarding non-proctored testing.

> Maybe it's because I'm young, but I'm aware of all these companies that students can pay to complete their assignments or even have them take their exams for them. There are many of them out there, so I was afraid my students would just pay someone to take their online exam for them.

However, a colleague provided some assurances to Professor Brown.

> He told me that those sites would be slammed during the pandemic. Everyone would be trying to get people to take tests for them. I actually went on one of those sites and posed as a student, and the wait time to take a test was like two weeks, so yeah not happening.

With the Coronavirus still strong, community colleges needed to keep most classes virtual during the 2020–2021 academic year, and this meant that testing continued to be remote. All five institutions employed testing where faculty could proctor students taking an exam via a web camera. Sisco, Habyan, and Griffin Community Colleges employed ProctorU. Telford and Bordi Community Colleges utilized Respondus, which locks down a student's computer, so they cannot browse the Internet while taking an exam. Faculty also had the option to meet with their classes live via Zoom and watch them take their exams to ensure there is no cheating.

> It's not perfect, but it's better. At least we can see that it is the student taking the test and make sure they aren't doing anything illegal, stated Professor Hickey.

> There were still logistical and technical issues. With Respondus there are all kinds of technical issues like Internet speed or connections, and with any method there are issues with students' web cams. It's just the best we could have done in this situation. I can't wait to get back to in-person testing. That's the way math testing always has been; that's the way it needs to stay forever. That's the way we can ensure fair testing, asserted Professor DeSilva.

The faculty also discussed how remote testing can create more anxiety for students.

> For some students, it's the camera. The camera goes on; they freak out. They feel like someone is watching their every move, which technically I am, shared Professor Williamson. Students are often taking these tests at home, and there can be a ton of distractions. For parents, their kids could be at home and could need something or the dog starts barking. When they are in the classroom, they can focus 100% on the test. They can't always do that at home.

What about Arithmetic?

Again, standalone arithmetic classes greatly decreased in the teens due to financial aid restrictions. Some schools offered short arithmetic refresher courses while some provided outreach for students. Such outreach programs included individual tutoring for students or referrals to adult basic educational programs for students who were very deficient in arithmetic skills. How were the students who were deficient in arithmetic served during the pandemic? The faculty at Sisco, Telford, Habyan, and Bordi Community Colleges continued to offer short-term arithmetic refresher courses and felt that these classes improved during the pandemic. Professor Mesa elaborated:

> You have to remember. In those classes, students have to self-pay. It's not like before when so many unmotivated people were taking the arithmetic classes. By and large, those students really want to improve their arithmetic skills so they can get better. The thing is; I've noticed that they can get embarrassed in class. They know how deficient they are in math, and that embarrasses them.

However, a virtual environment helped these students. Professor Rhodes explained:

> There seems to be more anonymity or maybe students are more comfortable asking questions on Zoom. So, I feel like they got better instruction.

Professor Hickey added:

> They seemed to work better in the virtual breakout rooms as well. I would put them into groups, and they would interact with each other very well, even more than in my higher-level classes.

Professor Moyer offered an explanation:

> The motivation is higher for these students. They have to do well enough in the arithmetic class to even take the basic algebra class.

Nevertheless, the pandemic hurt some students who needed arithmetic review:

> We try to work one-on-one with students in our math tutorial who need an arithmetic review, and we couldn't do that during the pandemic, stated *Professor DeSilva*. We were just scrambling to do everything virtually and online and we just didn't have the resources to help those students.

The students who relied on Adult Basic Education lost out as well.

> The ABE programs basically shut down due to COVID. I'll be honest with all the craziness, we were just trying to accommodate to life in the pandemic. We just didn't have the resources to help those students. So, I would say that many students who needed arithmetic help simply didn't get it, *recalled Professor Hickey*. I can't quantify how many students lost out. I just know we generally recommend a significant number of students each year to our ABE program, and we couldn't do that during the pandemic.

What Worked Well with Virtual Instruction?

While virtual instruction provided challenges for faculty and students, there were positive aspects to the modality.

Higher Comfort Level and Less Anxiety

Faculty had difficulty keeping students engaged in virtual classes; however, the faculty mentioned that several students felt more comfortable in a virtual math class.

> I had many students tell me that they felt so much more comfortable asking questions in class. In the past, they had been afraid to ask questions in class. They were just so afraid of looking stupid. Some even said that they would get so nervous, they couldn't even form the question correctly, said Professor Williamson.

Students loved the chat feature on Zoom, especially that they could reply just to me. No one else had to see their question. I had several students in my statistics class tell me that they got questions answered that they wouldn't have in a face-to-face class and that made the difference for them passing the class, reflected Professor DeSilva.

For other students, virtual instruction reduced math anxiety.

Even though we know students come in with math anxiety, I don't think we realize how severe it is, *asserted Professor Hickey*. For many students the anxiety is so severe, they have trouble concentrating and functioning in class. I had a lot of students tell me that being able to do math from the comfort of their home, as opposed to a classroom, put them at ease.

Professor Holton agreed:

I had a student tell me she had tried an algebra class once before but just couldn't focus in class because she was so nervous. She had such bad experiences in math. She said she had this fear of being class and just being lost and confused. She said it was just sitting in a class at a desk with the teacher at the board just brought all the anxiety back. But being able to take math in her home and in her room, she felt so comfortable.

While remote testing created anxiety for some students, it alleviated testing anxiety for others. Professor Rhodes elaborated:

Students tell me a lot of the test anxiety they get comes from anticipation of taking the test: being on campus, going to class, sitting at desk with the professor telling everyone to put everything away while he or she passes out the papers. Here, all they do is go on the computer and log on.

Professor Brown added:

Students get so nervous when they're taking the test and other students start gradually leaving when they are done. I tell them not to let that affect them, but I guess they can't help it. Anyway, the students tell me just the sound of chairs moving with other students getting up because they are done makes them lose their concentration. On Zoom, they don't even notice who is coming and going.

More Time on Task

The faculty mentioned that virtual instruction allows for more time on task, which can lead to more content coverage in class. Professor Smith explained:

Oftentimes in a face-to-face class, you get distractions. Students talk during class, and you have to stop and reprimand them. You'll get iPhones going off and you have to reprimand them. Then there is my ultimate pet peeve. Students start leaving early, and I have to remind them of when class ends.

Professor Harnich added:

> During the Zoom sessions, if a student gets disruptive with background noise, you can just mute them, and that's it.

Professor Sutcliffe discussed how distractions can impact how much content is taught in class.

> Sometimes when I have a [face-to-face] class that is getting really restless, I will wrap the class up, maybe 10 or 15 minutes early. I can just tell I that I have lost them, but in the Zoom classes, I just go till the very end of class because I can't even tell if they are restless. The benefit is I am getting more sample problems in. Maybe just me getting one or two more sample problems in helps some of them learn.

Recorded Sessions

Community colleges often have rules regarding videotaping class sessions:

> I've asked my supervisors about videotaping my classes. This way students who are absent can see what they missed, or really anyone can go back over things that we covered, but I have been told, I have to get signed permission from every student, and that just seemed too complicated, shared Professor Rhodes.

Fortunately, for the faculty and students, Zoom sessions can be recorded and posted online, and this has benefited students.

> In my evaluations, students have constantly stated that they love having links to the class sessions. In COVID times, students had to miss classes due to an emergency or a change in work hours, and it helped them to be able to see what they missed, asserted Professor Noles.

Professor Guzman added:

> Math is different from other subjects in that you can't just read about something and understand it. For concepts like asymptotes and logarithms and understanding trigonometric functions, you really need someone guiding you through it verbally from step 1 to step 2 to step 3 and so forth.

Professor Holton added that the recorded sessions also benefit those who were in class:

> I've had students tell me that they go back over questions that I covered in class just to get a better understanding. They say it's better than the YouTube videos because it's me teaching it, not someone they don't know.

Professor Smith concurred:

> It's impressive how many emails I get from students saying, "I was watching the class video and I have a question on this or that" and that student actually attended that class. They are simply going over the material to get a better understanding.

Success Rates for Virtual Instruction

Tables 8.1 and 8.2 compare the success rates (grades of A, B, C, or D) for each of the gatekeeper courses at Telford Community College from the Fall of 2019 (face-to-face instruction) to the Fall of 2020 (virtual instruction due to the pandemic). This includes both traditional lecture, emporium classes, and hybrid classes.

Explanations

While the success rates for the Fall of 2019 (in-person learning) and the Fall of 2020 (virtual learning) were somewhat comparable, the success rates were generally lower for the virtual learning period. I asked the faculty to provide their thoughts. Most attributed the lower success rates to higher attrition rates:

> We had more students dropping out of classes pretty much across the board, responded Professor Noles.

Why was this?

> I think a variety of reasons. It's just so much harder to engage students in virtual learning, *said Professor Stephens*. I think because they are not engaged, they lose interest, and any math course is rigorous, so if you're not engaged you can fall behind.

TABLE 8.1

Telford Community College Fall of 2019 Success Rates

Course	Pass Rate
Basic Mathematics (1 week)	62.5%
Introduction to Algebra (8 weeks)	52.7%
Elementary Algebra (8 weeks)	56.9%
Intermediate Algebra (8 weeks)	55.1%
Quantitative Reasoning	68.2%
Introduction to Statistics	64.1%
College Algebra	50.2%

TABLE 8.2

Telford Community College Fall of 2020 Success Rates

Course	Pass Rate
Basic Mathematics (1 week)	63%
Introduction to Algebra (8 weeks)	50.1%
Elementary Algebra (8 weeks)	55%
Intermediate Algebra (8 weeks)	53%
Quantitative Reasoning	67.9%
Introduction to Statistics	64.2%
College Algebra	48.6%

Some students get a little overwhelmed with virtual learning. I had a student tell me she "felt like she was on YouTube" during class, and that wasn't good because she felt a disconnect with the class, shared Professor Moyer.

Unfortunately, virtual learning means technical issues. Students have trouble accessing the class and maybe their computers don't work that well, so they have trouble even accessing the recorded videos, *stated Professor Rhodes.* I've had students tell me after they miss class, they try to watch the videos, but the videos keep stopping and loading. I'm no computer genius, but I'm pretty sure that means there is an issue with the student's individual computer.

The Style and Role of Virtual Classes

In general, the faculty asserted that they did not run their virtual classes all that differently from face-to-face classes.

I did what I normally do. Using the document camera, I showed some sample problems. I showed my PowerPoint notes. I did my best to engage them by having them work on problems, reflected Professor Williamson.

Obviously, there are things that I couldn't do under normal circumstances like walk around, talk to students, watch what they're doing, but I can still show problems on the white board using the document camera. I can access technology like Desmos. I can access different sites online. So, from an instructional point of view, I did what I normally do, stated Professor Brown.

I guess I still teach the same, but I can't engage and oversee as much, so it's on them to carry it out more, *shared Professor Guzman.* I spend time, especially in statistics, showing students how to use the scientific calculator. In the virtual classes, I can use the document camera to show them how to navigate the calculator, but I can't see how they're doing on the calculator. I can't walk around and watch them. It's on them to understand it and ask for help if they need it.

I asked the faculty if virtual classes in community college math have their place in post- pandemic life. All believe that they do.

It's like online learning. It's not for everyone. I think there are certain students whom this modality would work for, responded Professor Smith.

I look at virtual learning as halfway between traditional face-to-face and fully online. These are students who want and need the guided practice from their instructors and can ask questions and work synchronously with the class, but they need to be able to work independently to a certain degree, opined Professor Sutcliffe.

Just like online learning, students need to be technologically competent, *stated Professor Noles*. I had students who were intimidated by virtual instruction because the modality is so reliant on technology.

Virtual Instruction Can Increase Accessibility for Students

There have been positive outcomes to virtual instruction in community college math classes; therefore, this modality will likely continue to be a part of the post-pandemic world. In addition to in-class instruction, remote learning can assist students in other ways.

Instructor Absences

While math instructors should try to meet all their scheduled classes, emergencies and illnesses happen. Departments will likely attempt to find a substitute for an absent professor; however, this may not happen, and the class will be cancelled.

I'm in charge of finding a substitute when an instructor needs to be absent, *shared Professor Stephens*. And I can tell you too many classes get cancelled. This is especially the case when the instructor needs to be absent at the last minute.

Professor Harnich added:

Some students get annoyed when class is cancelled. They're annoyed that they came to campus and there was no class, but they're also upset about missing out on stuff. This is especially the case for students who are paying for their education themselves.

Professor Douglass discussed how even one absence can hinder student learning.

> I used to teach high school, and if you miss a day here or there, it doesn't make too much of difference, because you see the class like 180 times each year, but college is different. I had to miss a day of college algebra, and we couldn't find a sub [substitute], and I was just never able to catch up. I had to rush through the rest of the semester, and I don't think they got the best instruction. I only see them twice a week for 16 weeks, so missing one class makes a big difference.

The faculty spoke as to how virtual classes can resolve this issue going forward.

> If an instructor's going to be out, and we can't find coverage, they can record the lesson on Zoom and post it for the class, *suggested Professor Stephens.* I know it's not as good as everyone synchronously doing the material, but it's better than missing class. Students can watch it at their leisure.

Professor Douglass opined:

> Heck, sometimes, an instructor can't attend class but can still teach the class remotely with everyone following along. I mean if one of my kids gets sick, I can still teach the class from home. I just can't go to campus, because I have to stay home with my kid.

Virtual Tutoring

When the pandemic hit in 2020, college campuses had to shut down, and this included any person-to-person contact with students. Consequently, community colleges' tutorial services needed to find alternate methods to assist students with the daunting discipline of math. Like instruction, many tutorial centers utilized remote interaction. Professor Brown explained:

> Students would click on Zoom, and we had several tutors standing by. They could share their screens and show the tutors their questions. The students really appreciated getting help.

As community colleges allowed more students to return to campus for in-person classes, the tutorial services continued the remote option. Professor Harnich elaborated:

> Students just really liked the virtual option. Some were even telling us that they used our tutorial services more with the virtual option than they had in the past. It's just so much more convenient for them.

Professor Stephens added:

> I never really thought about this, but when a student has only one or two questions, they are much more likely to use the remote services than come to our in-person tutorial center.

Professor Douglass, who works with Sisco Community College's tutorial services as a faculty liaison, explained the challenge of offering both in-person and remote tutoring:

> You need a lot of support and a lot of organization. You need enough tutors to serve the students in-person, but you need enough tutors to monitor Zoom, and they have to be organized and act quickly putting students in breakout rooms with tutors. It needs to be a well-oiled machine.

Summary

The Coronavirus pandemic was an unprecedented time. Social distancing, to save lives, meant a major reduction for in-person learning, which led to virtual instruction. Consequently, faculty and students, very quickly, discovered the advantages and disadvantages of virtual instruction.

Virtual learning has some challenges. Many faculty members feel remote testing has its flaws. It is also difficult to engage students in a virtual setting. This can lead students to lose interest or fall behind in class. A virtual setting makes it more difficult for faculty to assess their students' progress on an incremental basis.

On the other hand, virtual learning can be advantageous. This modality can alleviate math anxiety, as students feel more comfortable working from their homes. Additionally, some students feel more at ease asking questions in a remote class, especially utilizing the chat feature. Video conferencing sessions, such as Zoom, allow the instructor to easily record and post the sessions, which can greatly help students. Ultimately, virtual learning can be beneficial to certain students and has its place in community college math education.

Reference

Online Learning Consortium. (2018). *2018 Annual Report*. https://olc-wordpress-assets. s3.amazonaws.com/uploads/2019/04/OLC-2018-Annual-Report-Online.pdf.

9

Learning from the Past and Present

The first eight chapters of this book have examined the history of gatekeeper community college math classes from the 1970s through the pandemic of 2020 and 2021. Additionally, we have studied math in higher education prior to the establishment of community colleges (Colonial Times through the twentieth century). What lessons can the past and present teach us?

Standalone Developmental Math Will Always Be a Part of Community College Math

The concept of developmental education has existed since the inception of American higher education, as tutors were needed to assist college students with courses instructed in Latin and eventually other subjects. This paralleled the idea of twenty-first century corequisite learning. Even though colleges served an elite student population, by the nineteenth century, this model, by itself, proved to be insufficient. That is why institutions began implementing standalone developmental math classes. To attempt to eliminate the need for developmental education, colleges began employing an admissions exam. Students still reported to college underprepared, and the need for developmental math persisted. In summation, long before the establishment of the open access community college, there was a need for standalone developmental classes and attempts to eliminate them.

As open access community colleges grew, the need for developmental math increased as well. Throughout most of the twentieth century, developmental math was accepted by college administrators, as colleges sought to help underprepared and underserved students. However, during the twenty-first century, developmental math courses have been redesigned and compressed in an attempt to increase student completion. While the redesigned pathways have value, the push to eliminate standalone developmental math is alarming.

Because community colleges are open access, students with very low reading and math skills may enroll. Default placement into college-level courses mean that these students will generally be overwhelmed by the material. The reality is many students need to build a basic mathematical foundation and even develop mathematical maturity before attempting a college-level class, even one with a booster component. It is possible that these students, because of their low-level skills, may not even succeed in a standalone developmental math class, such as introduction to algebra. However, these students are being denied other pathways such as arithmetic refreshers and even adult basic education programs to meet their needs.

Default placement into college algebra with a booster course can be especially disadvantageous to students. College algebra courses typically include difficult and sophisticated content such as polynomial and logarithmic functions as well as asymptotes, and conic sections. Mastery of this material requires a deep understanding of elementary and intermediate algebraic concepts. It is unreasonable to expect lower and medium level students to absorb such concepts and apply them immediately to college algebra concepts. Attempting college algebra without an understanding of intermediate algebra is unrealistic. While intermediate algebra is considered remedial by the states, the content in this course is challenging. Concepts such as quadratic equations, complex numbers, and system of equations require solid organizational skills and a strong aptitude of prerequisite skills. Students likely need to practice such skills before mastering them. In summation, the existence of college algebra, alone, requires the continuing need for developmental math.

The reality is that as long as open access community colleges exist, there will be a need for standalone developmental math classes. Eliminating such classes are a disservice to students. Unfortunately, there will be continued attempts to eliminate these classes for one reason or another. While these attempts will eventually prove futile, several students will be harmed in the process.

Full-Length Arithmetic Courses Are Ineffective, but Addressing Arithmetic Is Essential

Deficiency in arithmetic has been a longstanding issue in higher education dating back to the eighteenth century. Long before the existence of community colleges, college faculty have attempted to help students with arithmetic concepts. Understandably, standalone arithmetic classes became part of the community college math curriculum. During the 1970s and 1980s, faculty noticed that students were especially deficient in arithmetic and subsequently added additional arithmetic classes to the developmental math sequences. This allowed students more time to master arithmetic concepts and better apply them to algebra.

Full-length arithmetic classes became problematic in the aughts. Large numbers of students were testing into such classes, and few were succeeding. The combination of abysmal math skills, poor work habits, and various external issues led to low success rates. Arithmetic classes also overwhelmed developmental math and math departments. Students tested into these classes at such high rates that department chairs had difficulty covering these classes with faculty.

Success rates in arithmetic classes became so dreadful that the U.S. Department of Education restricted offering federal aid to courses that contained content below the ninth-grade level. This required community colleges to seek other ways to help students who are deficient in arithmetic. Ideally, there seem to be three effective pathways for these students, which are discussed in detail in Chapter 10.

Arithmetic skills are imperative to a solid mathematical foundation. While calculators cover arithmetic operations, students still need a conceptual understanding of concepts such as fractions, decimals, percentages and proportions for algebra. The development of number sense is imperative as mental math and basic math facts are necessary for higher-level math. In summation, college administrators must understand that students cannot attempt introduction to algebra or any equivalent course without a sufficient arithmetic background.

There Was Culture Change, Which Led to Culture Shock

The definition of culture tends to vary. However, culture as it relates to an educational organization is defined as the shared and fundamental beliefs that influence behavior (Schein, 1996). More specifically, a strong culture is one where there is a lot of unity in such beliefs; weak culture involves little unity (Kowalski, 2003). It may be odd to utilize the term culture within the discipline of math education. After all, culture tends to be associated with the social or behavioral sciences. However, when creating an environment for students to learn any discipline effectively, it is important to understand the school's culture and how this may impact learning.

When comparing community college math in the 1970s and 1980s to the late 1990s and especially the aughts, there was a noticeable change. Perhaps Professors Wallace, from Lester Community College, and Ballard, from Sisco Community College, summed it up best:

> When I first started teaching, the administration worked with us to help the students. By the time I retired, the administration was dictating to us what to do, asserted Professor Wallace.
>
> In the [19]70s and [19]80s, our administration supported us and helped us serve the students. By the time I left, we were running around ragged to help our administration look good and get money, reflected Professor Ballard.

The initiatives developed during the 1970s, 1980s, and even the 1990s were faculty driven, and they received support from college administrators. By the aughts, administrators were imposing top-down initiatives, some of which were ill-conceived, to increase student success. Additionally, several faculty members felt there were too many initiatives, and this stifled student success.

Another change worthy of note was the shift in mindset regarding developmental math. Throughout the 1970s and 1980s, developmental math was embraced. By the aughts, developmental math faced harsh criticism. College administrators were questioning teaching methods in developmental math and pushing for acceleration. Administrator's shared beliefs changed regarding both developmental math and a faculty member's pedagogical freedom. In the 1970s and 1980s, college administrators identified more with math faculty and saw them as peers rather than subordinates. Moreover, these administrators granted faculty more autonomy regarding pedagogical styles and supported them in implementing initiatives. However, by the 2000s, administrators were directing curriculum-related initiatives and even mandating pedagogical practices, something unheard of in the 1970s and 1980s.

This was a case of culture change. In their book, *Diagnosing and Changing Organizational Culture*, Cameron and Quinn (2006) discuss the four major types of organizational culture (Figure 9.1).

What caused the shift in culture? Finances played a major role. Toward the end of the twentieth century, state funding decreased across the country for public higher education. Consequently, community colleges became under increased pressure to raise student success rates, an issue that did not receive attention in the 1970s and 1980s. While colleges always competed for enrollment, this intensified for community colleges in the twenty-first century.

The Hierarchy Culture	The Market Culture
This organization focuses on stability and control. People are governed by procedures, formal rules, and polices. An organization's success is determined by dependability, low cost, and positive results.	Like the hierarchy culture, this is a results-oriented organization. People are competitive and goal oriented. Leaders are tough and demanding. Success is defined by beating out the competition and generating as much money as possible.
The Clan Culture	The Adhocracy Culture
The organization is considered a friendly place to work. People support each other, and leaders are more like peers and mentors. High morale is important in this organization, and teamwork is emphasized. Success is defined by sensitivity to customers and people.	This organization promotes individual freedom and innovation. Employees are encouraged to take risks and think outside the box. Success is defined by gaining new and unique perspectives.

FIGURE 9.1
Four types of organizational culture.

Community college administrators feared losing students to for-profit institutions, which allow more flexibility for students and less general education classes. Ultimately, community college administrators gravitated from a clan and adhocracy culture to more of a hierarchical and market-based culture. Faculty, however, still embraced a clan and adhocracy culture. More specifically faculty enjoyed collaborating with each other in a non-competitive manner. Furthermore, faculty valued utilizing various techniques to help their students. Consequently, this created a sense of culture shock for the faculty.

Why is understanding culture important? As seen in the aughts, culture clash between faculty and administrators can lead to poor morale and tumultuous times. Administrators were pushing reform initiatives to reduce costs, push students through developmental math, and help the community college seem superior to the competition. Faculty were focused on serving students in the best possible way. Consequently, some initiatives went awry, and this harmed student success.

While community college administrators are under pressure to reduce costs and increase student success, they may want to consider implementing a cultural audit (Whitt, 1993) when considering a new initiative. A cultural audit will study shared beliefs regarding a potential initiative and more importantly will give administrators an idea as to whether this initiative is a proper fit. There are various ways to conduct a cultural audit. For example, college administrators can conduct interviews or focus groups with math faculty and students to gain a better understanding as to whether this initiative will benefit students. In the aughts, learning communities, service learning, and inquiry-based instruction were flawed initiatives that created more harm than good for students. In the teens, the elimination of standalone developmental math proved disadvantageous to some students. In all cases, a cultural audit may have helped administrators to foresee some problematic issues. As long as community colleges are pressured to reduce costs and increase student completion rates and as long as community colleges must compete with for-profit institutions for enrollment, this kind of culture clash will exist. However, math faculty and administrators must work together to ensure that student learning is not thwarted in the process.

But It Has Always Been and Will Always Be about Money

While all schools are different, they have one characteristic in common. All schools raise revenue, and all schools spend money. The key is to make more revenue than is spent. Weisbrod et al. (2008) refer to this as the two-good framework. Cost factors drove much of the community college math reform movement throughout the twenty-first century. However, cost impacting math education was nothing new in American higher education.

Money has been impacting community college math and developmental math throughout the history of higher education. For example, in the 1970s and 1980s, developmental math was generously funded due to a combination of the discipline's high enrollment and consequential FTE production. Subsequently, developmental math grew, as community colleges added more courses and resources. However, once developmental math came under scrutiny and funding became tighter, community college administrators pushed for compression and even elimination of developmental math. Additionally, the need for revenue and concerns over the economy allowed higher education institutions to become less selective over the years. Without a doubt, cost will continue to factor into community college math education. However, as several faculty members advocated, the goal is to ensure that cost does not prioritize quality.

Alternative Math Pathways Have Been the Most Successful Initiative Thus Far

Throughout the twentieth century, community college faculty worked feverishly to develop resources and strategies to help students, especially underprepared students. In the twenty-first century, community college math education witnessed a great deal of reform initiatives to increase student completion. Throughout the history of community college math, some initiatives have succeeded; some have failed. However, none have been as successful as alternative math pathways for non-STEM students.

Creating shorter and alternative pathways leading to quantitative reasoning (QR) and introduction to statistics have benefited students greatly. More students are completing their required math sequences at a quicker pace. Additionally, less students are drowning in long developmental math sequences while enroute to college algebra, an irrelevant course for non-STEM majors. While corequisites are not a fit for all students, they have proven advantageous. Students can learn remedial material and very quickly apply such content to a college-level course without waiting a full semester. Ultimately, this means better completion, which leads to better funding and potentially more college graduates, which benefits society.

In addition to better success rates, these pathways have allowed students to make more meaning from their mathematical experiences. While algebra courses are imperative for students who need college algebra, it is difficult for students to relate concepts such as factoring expressions, simplifying rational expressions, and division of polynomials to real life. Faculty who teach QR courses reported that students are more engaged, as they can relate to concepts such as debt-to-income ratio problems as well as probability. Students also enjoy relating histograms, dot plots, and box plots to real-life situations.

This has also been the case for introduction to statistics, which covers concepts such hypothesis testing and confidence intervals, which are valuable to the behavioral and social sciences. Ultimately, these pathways allow students to apply math to their other coursework, their future careers, and their lives.

We Take Good Practices and Go Too Far

There have been many positive initiatives in community college math education. However, one fact is clear. We take positive and productive initiatives and practices to unnecessary levels. There are three major examples of this in the history of community college math: learning communities, QR, and corequisite education.

Community colleges utilized learning communities in the 1970s and 1980s to allow students to develop a sense of community and belonging by enrolling in multiple classes with the same group of students. This proved advantageous, as students became more comfortable with each other. Instructors even believed this helped them better master their math concepts. The modality resurfaced in the aughts with the same intention but had different results.

When learning communities reemerged in the aughts, the concept was not as simple. Students had to master more abstract concepts such as connecting common themes between classes. This proved difficult for students, and it took time away from their mastering the required math concepts. Moreover, and unlike the 1970s and 1980s, faculty felt micromanaged when teaching in a learning community. Faculty were required to employ more of an inquiry or group-based instruction, which led to confusion and frustration for students. The learning communities modality floundered because a primitive concept, such as allowing students to make connections with each other, became complex, abstract, and frustrating.

QR has positively impacted community college math. Furthermore, collaborative learning, which is part of QR, can help students better understand the content. However, college administrators and others have taken this concept too far. Organizations such as the Dana Center and Carnegie had heavy input when designing the QR course, and both advocated for group-based instruction being the primary source of instruction to provoke critical thinking. This has proven to be confusing and frustrating for students. QR is for non-STEM majors, and these students have struggled in and continue to struggle in math. QR also contains sophisticated concepts such as data description, finances, and compound interest problems. Consequently, many students need some sort of initial guidance from their instructors. Students must develop the skill before they can apply such skill.

In 2021, Complete College America (CCA) published *No Room for Doubt: Moving Corequisite Support from Idea to Imperative*. In this report, CCA discussed the failure of long developmental course sequences and the benefits of corequisite support for college-level math classes. CCA also suggested default placement into college-level classes, which means the removal of standalone developmental math classes. Additionally, CCA urged community colleges to act quickly.

Some students can assimilate new content in a booster course and then quickly apply it to a college-level course. For these students, corequisites are beneficial and can help them complete their math requirements at a quicker pace. However, *No Room for Doubt* fails to explain how booster courses can serve students who simply need more time to assimilate prerequisite content or are severely lacking in prerequisite skills. Booster courses do not have the bandwidth to serve as a remedy for all underprepared students. Eliminating standalone developmental math and relying completely on corequisites is unfair to students who need to build a foundation before attempting a college-level math course.

The Emporium Model Will Always Be Around in Some Form

The emporium model of instruction in developmental math has existed since the 1970s. Some faculty were overwhelmed with the large and heterogeneous group of students, and they began employing the emporium model, which allowed students to work at their own pace and also allowed instructors to work with students on a one-to-one basis.

The concept of the emporium model has remained the same. It consists of students working at their own pace in a lab setting. However, its characteristics have changed over the years. The emporium model began with students completing worksheets, which evolved to VHS videos, then to Apple Computers to CD ROMs, and finally to interactive math software networked to the instructor.

The emporium model became a major part of the reform movement of community college math in the twenty-first century. The modality has received mixed reviews. Some praise the individual instruction that students receive and the opportunity to accelerate through their developmental math sequences. Others assert that the emporium model is a poor fit for students, as many students need guided practice from their instructors before attempting math problems on their own. Consequently, some students who attempt math in the emporium model become baffled and frustrated. Additionally, some mathematical content can be more difficult in the emporium model. Some faculty mentioned that concepts in higher-level algebra, such as system of equations and quadratic equations, can be more challenging for students in the emporium model, as opposed to lower-level algebra concepts. This is because such higher-level concepts require more organizational skills and more of a mathematical foundation.

The emporium model has its place in community college math. Some students learn better in a lab setting and benefit from self-paced instruction and drill and practice. The dangerous aspect of the emporium model is the cost and the political factor. For example, it is cheaper to run a class in a lab with 50, or more, students with one instructor and one or two tutors, as opposed to a traditional class of 30 with an instructor. This is dangerous, as college administrators may want to extrapolate the emporium model across developmental math programs and eliminate traditional instruction by prioritizing cost over quality. Additionally, Hanson (2003) defined politics as the competition for resources. College administrators engage in politics to obtain funding. Colleges have implemented various learning initiatives so that the institutions can appease state legislatures or justify or obtain funding. The emporium model is an example. However, when implementing such initiatives, quality must be placed over cost, and pedagogy must be prioritized over politics.

Will Online and Virtual Instruction Replace In-Person Learning?

Distance learning became popular in community college math as a form of correspondence learning in the 1990s. However, enrollment in the modality surged in the late 1990s and continued growing in the 2000s, especially as distance learning morphed into online learning. Math software programs that are networked to the instructor and allow for interactive learning have only increased the popularity for online learning. During the pandemic, the discipline of community college math was introduced to virtual instruction, which allows for students to synchronously learn math with their classmates, participate, and ask questions from a remote location. Students also make connections with each other and form a sense of community that is difficult to replicate in an online environment. There is no question that both online and virtual instruction are a major part of the future of community college math.

With technology constantly evolving and the popularity of online and virtual instruction growing, will in-person instruction be phased out of community college math? This is very unlikely. In-person learning is a necessity, for some students, when mastering math. The combination of face-to-face instruction and synchronously learning with classmates is imperative to student learning. In-person learning also allows an instructor to monitor and assess students' progress incrementally. This modality permits instructors to work with and correct students' organizational habits, an imperative part of math education. In summation, technology will evolve to unimaginable lengths in the future. It is incomprehensible to imagine what our technological capabilities will be years or decades into the future. However, in-person learning will always have a vital place in community college math.

Effective Teaching Involves Well-Rounded Instruction

How does a faculty member teach math effectively? There is no easy answer. Every student is different, and every faculty member is different. Teachers have their individual comfort levels as well as strengths. However, there seem to be some common threads regarding effective math instruction.

Ambiguity in math frustrates students. This leads to math anxiety and an overall disdain for math. Consequently, students, especially in gatekeeper math classes, need thorough instruction. When introducing a new concept, instructors must explain this concept in detail from beginning to middle to end. Several instructors asserted how this requires preparation on their part. This involves instructors completing sample problems in advance and even practicing how they will teach the material.

Engagement is an imperative part of effective teaching. More specifically, instructors must find a way to engage students in their own learning. This may involve asking students questions as they explain a topic. It may involve allowing students to work through problems in groups. This also may involve games or activities in the classroom that help students master the content.

Effective teachers are also able to assess students' needs. This does not simply mean assessing whether students understand the material after giving an exam. Effective instructors find ways to embed assessment into their daily class lessons. This may involve asking students questions during a guided practice or employing brief written assessments in class. More importantly, such assessments should guide the instructor's lesson. This could inform instructors whether to spend more or less time on a topic or even utilize a different approach.

Math wars have emerged in both K–12 and community college education over effective pedagogical practices. Progressives assert that math should be taught employing group-based instruction or inquiry-based instruction. This leads to critical thinking and will allow students to master the content thoroughly. However, several instructors asserted that while collaborative learning can be advantageous, students need a basic foundational skillset before application. Students who enroll in community college with math anxiety and a low math skillset require and desire thorough instruction. Taking students by the hand and guiding them helps build their knowledgebase and eases their anxiety. In fact, for community college students, learning math parallels Bloom's Taxonomy (see Figure 9.2)

Students must remember and understand math facts and formulas before moving to higher order thinking such as applications and analysis. Additionally, students who continue to develop a solid mathematical foundation can evaluate their own work. Some faculty noted that highly developed students can even create their own math problems.

Technology has evolved greatly since the 1970s and it will continue to evolve over time. It is imperative that instructors stay current with technology.

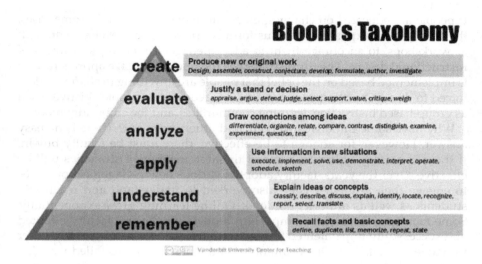

Bloom's Taxonomy

create — Produce new or original work
Design, assemble, construct, conjecture, develop, formulate, author, investigate

evaluate — Justify a stand or decision
appraise, argue, defend, judge, select, support, value, critique, weigh

analyze — Draw connections among ideas
differentiate, organize, relate, compare, contrast, distinguish, examine, experiment, question, test

apply — Use information in new situations
execute, implement, solve, use, demonstrate, interpret, operate, schedule, sketch

understand — Explain ideas or concepts
classify, describe, discuss, explain, identify, locate, recognize, report, select, translate

remember — Recall facts and basic concepts
define, duplicate, list, memorize, repeat, state

Vanderbilt University Center for Teaching

FIGURE 9.2
Bloom's Taxonomy.

Students will likely enter community college with current technological abilities; therefore, instructors should be able to embed current technology into their classes. However, it is noteworthy that the use of technology by itself does not lead to effective teaching. It can only supplement effective teaching. The use of thorough instruction, student engagement, and effective assessment have always constituted effective instruction and always will.

Change Will Never Occur Unless There Is Legitimate Faculty Buy-In

Community college faculty have faced various changes over the years. Some changes have effectively improved instruction while other changes were ineffective and faded. For an academic discipline to be effective, change is necessary over time. However, unless faculty buy into new initiatives and new ideas, change will not occur.

According to Chin and Benne (1985), administrators will utilize three strategies when attempting to implement organizational change. The first strategy is the empirical-rational approach. This is based on the belief that when shown data or evidence, people will act rationally and implement such change. Over the years administrators and others have shown faculty evidence of poor success rates for a variety of students. After viewing such evidence, administrators believed that faculty would see the need to make pedagogical changes. The second strategy is the normative-reeducative approach.

If people are educated on changes, they will implement them. In some cases, the faculty were exposed to various forms of professional development, such as workshops, to encourage them to adopt pedagogical changes or forms of instructional design. The third strategy is the power-coercive approach. This is implemented based on the belief that people are unwilling to change unless forced to do so. Over the years, faculty were required to adopt initiatives such as group-based instruction, learning communities, and the emporium model.

Why have some initiatives worked, and others have not? There is no easy answer. However, for change to be effective, there must be faculty buy-in. If there is no faculty buy-in, none of the aforementioned strategies will be effective. Over the years, faculty adopted various new pedagogical practices to engage their students. They also worked to understand and reach ESL students as well as minority students. Online learning grew exponentially. Some faculty embraced the emporium model, and several faculty adopted and successfully implemented alternative math pathways. Conversely, other initiatives have failed. One common thread when initiatives failed was that there was no faculty buy-in.

How can administrators and others obtain faculty buy-in when implementing change? Moreover, are most faculty resistant to change? Change can be intimidating, but it is important to remember that there have been many faculty-driven initiatives that have led to positive outcomes. Earlier in this chapter, I discussed the concept of a cultural audit. Again, this is one way administrators can determine if a potential initiative is a good fit. However, there is no way to force faculty to accept and implement instructional change. One consistency among faculty through the years is their devotion and care for their students' learning. If faculty do not ultimately feel that a pedagogical practice or learning modality is enhancing their students' learning, they will withdraw or alter that practice. Moreover, if faculty feel that pedagogy is being comprised by politics, they may become resistant to change.

There Is No End in Sight for Student Entitlement and Student Consumerism

As the twenty-first century has progressed, there has been an increase in both student entitlement and student consumerism. This has particularly been the case for Generation Z students as well as the latter part of Generation Y students (Harrison & Risler, 2015). An example of student entitlement involves expecting a passing grade simply for not legitimately passing the class. While the corequisite model has been effective, this can be an another example of student consumerism as this is accommodating the students' desire of a shorter math sequence with less developmental math. Eliminating developmental education is a clear case of student consumerism. Students rejecting the adult basic education classes because they will not receive financial aid as well as administrators discouraging such classes out of fear of losing students is a mixture of both entitlement and consumerism.

Students are entering community college with a heightened sense of entitlement. Additionally, community college administrators have implemented various initiatives to retain students and compete with other institutions. Two facts are clear going forward: Student consumerism can negatively impact the learning of math, and student consumerism does not appear to be dissipating any time soon.

While Times Have Changed, Math Content Has Remained Stagnant

Through the years, community college math education has evolved. There have been technological advances. Initiatives to improve student success have come and gone, and in some cases, come again. We have been introduced to teaching and classroom modalities such as the emporium model, online learning, and even virtual instruction. Course names have changed. However, through it all, math content has generally remained the same.

When comparing math content from the 1970 to the early 2020s, there is very little difference. Certainly, schools have shifted content from one course to another. There are some topics such as the binomial theorem, which are not taught as much, at least in the traditional sense. Conversely, debt-to-income ratio problems were not emphasized as much in the past, but this topic became more popular with the growth of the QR course, which emphasizes real-world applications. However, in general, topics in algebra classes have consisted of operations with rational numbers, linear equations quadratic equations, roots and radicals, and complex numbers. College algebra has emphasized functions, logarithms, and sequences and series. Statistics courses have focused on data description, probability, hypothesis testing, and permutations and combinations.

Why has community college math content stayed relatively stable? Look no further than calculus. To be successful in calculus, students need to master all aspects of algebra. As Professor Smith, from Griffin Community College, noted:

> I always have to tell students that roughly 70% of most calculus problems are based on algebra. So, if you don't understand concepts such as factoring, quadratic equations, or rational expressions, you won't be able to simplify the calculus application.

Consequently, there must be algebra courses as well as a college algebra course that addresses these prerequisite skills. The same can be said regarding statistics. Professor Noles, from Telford Community College, opined:

> Most students who take introduction to statistics aren't going to be statistics majors, but some do take more statistics classes, and they need to know about factorials, probability, and confidence intervals, so by default, that content will always be in a basic statistics class.

Another reason math content has remained rather stagnant is that math, by design, is a very linear and progressive discipline. This is a concept that some non-math individuals do not seem to understand, which has been frustrating to math faculty. Professor Mussina explained:

> When our administration has wanted us to compress our courses or redesign our curriculum, I feel like we have to remind them that math is linear and progressive. If you don't get factoring, you won't get rational expressions. If you don't understand the product and quotient rules of exponents, you won't be able to compute exponents with combined rules or negative exponents. If you don't understand rational expressions, you can't do concepts like rational inequalities or simplify trigonometric identities.

Professor Hickey added:

> I don't think a lot of administrators or so-called experts understand how linear math is and how important it is to master certain prerequisite concepts before tackling subsequent concepts. We were trying to explain the challenges of college algebra with the corequisite. We couldn't get them to understand basic facts such as students have to understand how to solve by completing the square so they can get circles and hyperbolas in standard form in college algebra.

The reality is that no amount of redesign or reform can alter the linear nature mathematics. Students learn at different paces; however, students must master prerequisite concepts before tackling future concepts. That has always been the case for most mathematical concepts and always will be.

Summary

The discipline of community college math has a detailed history. It is imperative to learn from both the past and present to plan for an effective future. While developmental math has become unpopular in the twenty-first century, this discipline provides a great deal of value for some students. Some students benefit from truncated refresher courses while others require full-length courses. Some students profit from an emporium model while others need face-to-face instruction. Flexibility is the key.

Alternative math pathways have positively impacted community college math. Allowing non-STEM majors to enroll in QR or introduction to statistics, as opposed to college algebra, has allowed students to complete their math requirements at a quicker pace and improved their overall experiences in math. Additionally, corequisites have helped students accelerate through their math courses as well. However, these pathways must be managed carefully. Math is a linear and progressive discipline. It is unforgiving for those who lack proper prerequisite skills. The pressure to reduce cost and engage in politics cannot prioritize effective pedagogy and students' needs.

There was a period in the late nineteenth century and early twentieth century when math was not required in higher education. It is unlikely this will happen again. Math has become too important in American education. Considering this, how can we design an effective math program for years to come to help our students succeed?

References

Armstrong, P. (2010). *Bloom's taxonomy*. Vanderbilt University for Teaching. https://cft.vanderbilt.edu/guides-sub-pages/blooms-taxonomy/.

Cameron, K. S. & Quinn, R. E. (2006). *Diagnosing and changing organizational culture: Based on the competing values framework*. Jossey-Bass.

Chin, R. & Benne, K. D. (1985). General strategies for effecting changes in human systems. In W. G. Bennis, K. D. Benne, & R. Chin (Eds). *The panning of change* (4th ed., pp. 22–43). Holt, Rinehart, & Winston.

Complete College America (2021). *No room for doubt: Moving corequisite support from idea to imperative*. https://completecollege.org/wp-content/uploads/2021/04/CCA_NoRoomForDoubt_CorequisiteSupport.pdf.

Hanson, E. M. (2003). *Educational administration and organizational behavior* (5th ed.). Pearson Education Inc.

Harrison, L. M., & Risler, L. (2015). The role consumerism plays in student learning. *Active Learning in Higher Education, 16*(1), 67–76. https://www.learntechlib.org/p/159326/.

Kowalski, T. J. (2003). *Public relations in schools* (3rd ed.). Prentice Hall.

Schein, E. H. (1996). Culture: The missing concept in organizational culture. *Administrative Science Quarterly, 41*(2), 229–240. doi: 10.2307/2393715.

Weisbrod, B. A., Ballou, J. P., & Asch, E. D. (2008). *Mission and money: Understanding the university*. Cambridge University Press.

Whitt, E. J. (1993). Making the familiar strange: Discovering culture. In G. Kuh (Ed.), *Cultural perspectives in student affairs work* (pp. 81–94). Lanham, MD: University Press/American College Personnel Association Series.

10

Suggested Mathematical Models for Sustaining Success

Mathematics is a discipline that students have long struggled with at community colleges. In reporting conversations with various faculty members from this study and examining both the quantitative and qualitative data gathered, this chapter presents suggested pathways to help a heterogeneous group of students navigate community college math for years to come.

Returning Participants

Additionally, for the concluding chapter, the following participants shared their experiences and views: Professors Rhodes (Telford Community College), Hickey (Habyan Community College), Smith (Griffin Community College), Harnich (Bordi Community College), Williamson (Habyan Community College), Moyer (Telford Community College), Noles (Telford Community College), DeSilva (Sisco Community College), Morgan (Habyan Community College), Milacki (Bordi Community College).

Suggested Pathways for Sustained Success

The tables below (Figures 10.1–10.4) summarize suggested pathways to sustaining student success for both non-STEM and STEM students. The pathways allow non-STEM students to complete their college-level math requirements within 1 year. All pathways minimize cost and time while also ensuring preparedness. Additionally, community colleges should consider offering introductory algebra as a 1-week, compressed course for advanced students, which would accelerate their progress.

DOI: 10.1201/9781003287254-10

Note: Students who have completed or tested out of an elementary or middle-level algebra class would not need the booster course with quantitative reasoning (QR).

FIGURE 10.1
Non-STEM students: liberal arts or fine arts majors.

Note: Students who have completed or tested out of an elementary or middle-level algebra class would not need the booster course with introduction to statistics.

FIGURE 10.2
Non-STEM students: behavioral or social science majors.

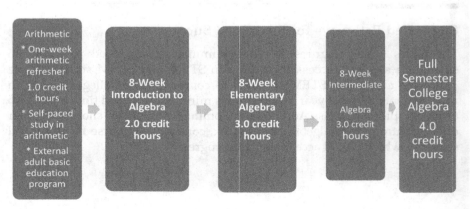

FIGURE 10.3
STEM majors option 1.

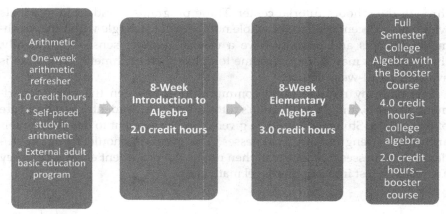

Below is the rationale for each class and pathway.

FIGURE 10.4
STEM majors option 2.

Arithmetic

Some students will enter community college deficient in arithmetic skills. This has always been the case, and it will never change. The question is, how can community colleges serve these students?

Pathway 1

For students who place at the higher end of a testing instrument in arithmetic skills, community colleges should offer a truncated—possibly a week-long—arithmetic refresher course. Such courses should meet several times during the scheduled week, amounting to, for example, 20 hours. These refresher courses could be offered face to face or online. Since these students should have a mastery of whole numbers, these courses can focus on operations with fractions, decimals, proportions, percentages, and word-problem applications. Ultimately, these are the essential topics required for basic algebra.

Pathway 2

Some community colleges have provided students with individualized programs to help students in arithmetic. This instruction is generally done through the school's tutorial center. These programs are ideal for students who have a decent mastery of whole numbers but struggle with other arithmetic concepts and possibly have a weaker number sense. Additionally, these students may want more time to polish their arithmetic skills than is available in a 1-week course.

This pathway requires solid communication between the math department, advisors, and the institution's tutorial services. Students must be aware of this option. Students should be given a pre-assessment to determine their arithmetic strengths and weaknesses. The program should focus on students' weaknesses. Students can then retake the placement exam when they are ready to test into a higher-level math class.

Pathway 3

Unfortunately, there are students who struggle with whole numbers and basic number sense. For these students, a state-funded adult education program is the best fit. Such programs allow students to develop their basic skills at their own pace. Community colleges should work with these programs to ensure that students' needs are met. While completing these programs, students should be in contact with academic advisors and other mentors to keep track of their progress.

Pre-Algebra/Introduction to Algebra

Before attempting a QR or statistics, even with a booster course, students should complete or test out of a pre-algebra class.

Target Group: This class focuses on students who possess an arithmetic foundation but very little or no algebra skills. Additionally, it serves students who are on either non-STEM or STEM pathways.

Content: Topics in the class should include:

- Operations with rational numbers
- Evaluating expressions
- Linear equations and basic geometric applications
- Literal expressions
- Word problem applications

Course Length: Pre-algebra could take the form of an 8-week (half-semester) class, allowing students who are unsuccessful during their first attempt to complete the course in the second half of the semester. However, pre-algebra could be offered as a full semester course as well. For students who simply need a refresher, this course could be offered in a truncated but intense 1-week format with roughly 20 hours of contact time. Again, this truncated format should be for higher level students.

Instructional Modality: Students benefit from face-to-face instruction, which should include engagement and connection with peers. The self-paced emporium model, with no lecture, is ideal for some students, as it allows them to work at their own pace and focus on their weaknesses. Online instruction should be reserved for higher level students who can learn independently.

Rationale: Clearly, students who are on the college algebra pathway require pre-algebra, but why is pre-algebra knowledge imperative for a QR or statistics class? Students need basic algebraic skills, such as evaluating expressions and solving equations, for these classes. More specifically, these skills come in handy when computing probability problems in both classes, as well as finance problems in the QR class. Being able to navigate literal expressions and applications with formulas will serve students as they contend with more complex formulas in both statistics and QR. While algebraic word problems, such as consecutive integer problems, do not translate to QR or statistics, students need to understand how to approach a word problem. More specifically, to be successful in a QR or statistics class, students must be able to successfully read mathematics.

Elementary Algebra

Target Group: This class is mostly for students on the college algebra track. It is also an option for students who do not wish to take the booster course for QR or statistics. However, data suggest that students with sufficient knowledge of pre-algebra should be able to complete QR or statistics with the corequisite.

Content: Topics can vary, but this class should include mid-level algebraic concepts, such as:

- Lines (slope and equation of the line)
- Factoring expressions
- Operations with rational expressions
- Complex fractions
- Rational equations
- Operations with polynomials
- Word problem applications

Course Length: Like pre-algebra, this class should take 8 weeks. This time-frame allows students to complete both pre-algebra and elementary algebra in one semester. The content in this course is likely too difficult to offer in a 1-week format.

Rationale: These topics are all prerequisites for intermediate algebra content.

Instructional Modality: A face-to-face option is imperative. The emporium model is an option, but students must be carefully placed. Content in this course is more difficult than pre-algebra, and the problems are much longer. For students who are learning, as opposed to reviewing, elementary algebra, the face-to-face option is best, as students need thorough guided practice with this more difficult content. Again, online learning serves more advanced students.

Intermediate Algebra

Target Group: Intermediate algebra is for students on the college algebra track. Additionally, this standalone course is ideal for students who lack an understanding of intermediate algebra and would be overwhelmed in a college algebra booster course.

Content: Intermediate algebra should include higher level algebra concepts, such as:

- System of linear equations [two-by-two and three-by-three]
- Roots and radicals
- Quadratic equations and parabolas
- Absolute value equations
- Word problem applications
- Operations with complex numbers

Course Length: An 8-week class is ideal. Students can retake it in the second half of the semester if they are unsuccessful without losing too much time. Like elementary algebra, the content in this course is too rigorous for a 1-week, truncated version.

Rationale: Mastery in these topics is imperative for being successful in courses featuring polynomial, rational, and logarithmic functions. These topics are also essential for working with various conic sections as well. For example, it is necessary to solve by completing the square when getting an ellipse or a hyperbola into standard form.

Instructional Modality: A face-to-face modality with thorough instruction and student engagement is essential. The emporium model is less effective. Intermediate algebra content is arduous, and students must develop painstaking organizational skills to be successful. Mastering these concepts is more challenging in a lab environment. However, schools that employ an emporium model may want to consider a lecture and lab format, where students engage in some synchronous guided practice.

Quantitative Reasoning

Target Group: This course is for non-STEM majors.
  Content: QR should include:

- Data description (dot plots, histograms, box plots)
- Interpretation of charts and graphs
- The empirical rule and probability applications
- Finance applications (debt-to-income ratios, simple and compound interest)
- Linear and exponential regressions

Course Length: This should be a full semester class. An 8-week QR course would allow students to complete pre-algebra and QR with the booster in one semester. However, the amount of content covered and time on task needed for students for QR with the booster in 8 weeks may overwhelm students.
  Rationale: The purpose of QR is to provide non-STEM majors with an understanding and contextualization of math in the modern world. The entire course, utilizing the aforementioned topics, should consist of real-life applications that relate math topics to the current time.
  Instructional Modality: This course is full of applications that require critical thinking. However, students must master basic skills before being able to contextualize them. For example, students must understand basic probability formulas and comprehend the difference between mutually exclusive and non-mutually exclusive events, as well as events connected by "and" or "or," before applying such skills correctly. Students need guided practice from their instructors; however, after they have developed this foundation, collaborative learning can provoke more critical thinking.

Quantitative Reasoning Booster Course

Target Group: A QR class with a booster course is suitable for students who can assimilate basic content, such as finding slope of a line and solving fractions with equations, and quickly apply it to college-level content.

TABLE 10.1

Quantitative Reasoning Booster Course Topics

Quantitative Reasoning Topic	Prerequisite Skill Required
Histograms	An understanding of how to obtain equivalent intervals from raw data
Box plots	Median
Empirical rule	Mean
Probability	Simplifying expressions with fractions and decimals using a calculator
Debt-to-income ratio problems	Solving equations with fractions
Simple interest problems	Finding the slope and equation of a line Evaluating expressions
Compound interest problems	Evaluating literal expressions with exponents
Linear regression	Finding slope and equation of a line
Exponential regression	Evaluating literal expressions with exponents.

More specifically, students must be able to handle being in a math class (QR with a booster) for roughly 4–5 hours per week while still completing the work for both classes.

Content: Some aspects of the QR course content do not require prerequisite skills. However, Table 10.1 lists topics that do.

Instructional Modality: Considering that students who need the booster course possess weaker skills than those who do not, much of the class should consist of extra review and practice. Instructors should spend time asking students if they have questions or providing the class with extra practice and application problems. These problems could be completed in groups to foster student engagement.

Introduction to Statistics

Target Group: Like the QR course, introductory statistics is suitable for non-STEM majors; this class is a good fit for behavioral science, social science, and some business majors.

Content:

- Measures of central tendency
- Data description (histograms, dot plots, box plots, stem and leaf plots, line graphs, and bar graphs)
- Computing variance and standard deviation
- Understanding Venn diagrams and set theory
- Probability (independent and dependent)

- Discrete probability
- The normal distribution
- The central limit theorem
- Confidence intervals
- Hypothesis testing
- Chi-square distribution and chi-square tests
- Linear regression equations
- Identifying bias

Modality: Traditionally, introductory statistics courses have been taught in traditional, face-to-face settings, and this tendency should continue in the future. Introduction to statistics is a fast-paced course that consists of rigorous content. Students need thorough and guided practice, and instructors reported that students benefit from hands-on activities in their statistics courses. Such activities deepen their conceptual understanding of the content.

Introduction to Statistics Booster Course

Target Group: Like the QR course, there is minimal algebra required for introductory statistics. However, some topics do require prerequisite skills from students. The table below shows which prerequisite topics should be covered in a booster class to prepare students for particular topics in the statistics class (Table 10.2).

TABLE 10.2

Introduction to Statistics Booster Course Topics

Introduction to Statistics Topic	Prerequisite Skills
Data description	Interpreting bar graphs and identifying equal intervals from a data set.
Probability	Sample space and using the calculator to simplify expressions with fractions
Discrete probability	Solving linear equations and linear inequalities with fractions
Confidence intervals	Plus–minus notation Evaluating expressions Equations with square roots
Hypothesis testing	Understanding the T-table Equations with square roots
Chi-square	Summation formula
Linear regression equations	Linear equations

Instructional Modality: Like the QR booster course, there is time to both prepare students for introduction to statistics and to review material. Additionally, since an introduction to statistics class relies heavily on the calculator (either scientific or graphing) instructors should spend time during the booster class focusing on the logistics and syntax of the calculator.

College Algebra

Target Group: Until the teens, most students needed to complete college algebra to gain credit for a transferrable college math class. Going forward, college algebra should be taken solely by students who require calculus.

Course Length: This should be a full-semester class with no truncated versions offered due to the rigorous content.

Content: This course should include:

- Polynomial, rational, radical, exponential, and logarithmic functions
- Graphical representations of functions
- Rational inequalities
- System of linear inequalities
- Matrices
- Conic sections (circles, ellipses, and hyperbolas).

College Algebra Booster Course

Target Group: Enrolling in college algebra with a booster course is not for all students. Students must be able to assimilate content from intermediate algebra and quickly apply that content to college algebra topics. Essentially, students must learn content from a difficult course and then quickly apply it to another difficult course.

Content: Students must master the following algebra topics in the college algebra booster course. Several of these topics must be applied to multiple college algebra topics:

- Expansion of binomials
- Finding slope and equation of the line
- Operations with polynomials (addition, subtraction, multiplication, and division)

- Factoring all types of polynomials
- Operations with rational expressions (addition, subtraction, multiplication, and division)
- Solving inequalities, quadratic equations (factoring, quadratic formula, square roots, and completing the square)
- Operations with complex numbers
- Solving basic radical equations
- System of linear equations

The aforementioned topics are a mixture of elementary and intermediate algebra content, depending on how each school arranges its courses. However, all are imperative for the mastery of college algebra topics.

Instructional Modality: Unlike the QR booster course, in the college algebra booster course, instructors will need to spend more time teaching remedial material to prepare students for college algebra, and there will be less time for review. More specifically, a college algebra booster course requires thorough instruction in intermediate algebra.

Logistics for Booster Courses

Departments need to address two important items regarding booster courses: the length and instructor for the course. More specifically, should the faculty member who teaches the college-level course also teach the booster course?

Length

I asked the faculty, who taught during the teens, how long a booster course should run. In general, the faculty asserted that 2 hours per week, which is equal to two semester hours is sufficient for a booster course.

> You have to keep in mind that they are pretty much taking two math classes at once. So, more than two hours for the booster class is too much, said Professor Rhodes.
> It's [the two hours] more of a challenge for the college algebra booster class, because you basically have to teach all of intermediate algebra, but it can be done, shared Professor Hickey.

Should the booster classes meet on the same days as the college-level class or on different days? All instructors agreed that booster classes should be offered directly before the college-level class.

> We tried the booster class on different days, *reported Professor Smith.* Our statistics booster classes met on Mondays and Wednesdays, and our regular [college-level] statistics classes met on Tuesdays and Thursdays. In theory it sounded good. It wasn't too much math in one day, and they could learn something [a prerequisite topic] on Monday, let it sink in, and then apply it Tuesday. It just didn't work with students' schedules. So many of them with their work schedules only want to be on campus two days a week.

Professor Guzman agreed:

> It's not ideal to have students in a math class for three or four hours a day, but that's the least of all the evils. We get them there for the booster course, give them a break, and then they take the regular [college] class. At first, we tried the booster classes on different days, but we would either have trouble getting them to register or they would register and miss the booster class.

Virtual Booster Classes?

Considering the scheduling issues and the ability to hold virtual classes, should math programs make the booster classes virtual and hold the college-level section in-person?

> I think that would be a mistake, *said Professor Harnich.* The thing you have to understand about the booster classes is that as the semester goes on, they drag. I don't know why that is, but after about seven or eight weeks, engaging the students in the booster class becomes hard. Some of them stop coming. So, given it's harder to engage students in the remote setting, I think the booster classes work better face-to-face.

Professor Williamson shared:

> During the pandemic, when we were all on Zoom, the booster classes were the worst to teach. The classes are small, and the students just don't engage. Usually during booster classes [face-to-face], I give them problems to do, put them to work, but mostly I walk around and talk with them and watch them do the problems. You just can't do that as well on Zoom. So, I would give them problems and go over them, but I felt like I was talking to myself. After a while, during the booster classes, I would just open up the Zoom room and let them drop in with questions. I got so tired of coming up with things to do and talking to myself.

Professor Moyer concurred and provided an offering as to why the booster courses tend to lag as the semester progresses:

> A college-level class is really fast paced. As the teacher, the adrenaline is always going. There is pressure to teach well and cover everything [the content] in the time allotted. The students also have to keep up and

stay alert, but in a booster class it's all remedial material and review, and that's why I think it drags. No, booster classes don't work well virtually. You'll lose even more students.

Instruction

Some schools have the same faculty member teach the booster class and the regular class, while others utilize different instructors. All the faculty agreed it is better to employ the same instructor for both.

> We use different teachers, and it can get really messy, *noted Professor Smith.* Sometimes the right hand doesn't know what the left hand is doing.
>
> People have to understand; the booster class is really touch and feel. Yes, there is content we have to cover to prepare students, but a lot of that class is going over questions they have and also focusing on weak parts. The instructor who teaches the college sections knows that best, asserted Professor Noles.

Enrollment Flexibility

The faculty also noted that there should be flexibility regarding late enrollment into booster courses. Professor Williamson explained:

> Sometimes a student completes a previous course or places out of a booster course, but they realize later on that they need the booster class.

Professor Williamson clarified:

> I have had students who take my college algebra class who either passed intermediate algebra or tested out of it. The problem is after a week or two, they realize they need more help with intermediate algebra. Maybe it's been a while [since they took intermediate algebra] or there are concepts they just struggle with.

Professor Noles concurred:

> I've had students in my college algebra class ask me two or three weeks into the semester if they can join my booster class, and I am fine with that. I think schools need to be flexible to help students.

Alternative Math Pathways Reduce Time and Cost for Students

While alternative math pathways ensure that students are prepared for their subsequent math classes, such pathways reduce the time and cost for students. If students place into introduction to algebra, they could complete their college-level math requirement (QR or statistics) within 1 year. Additionally, students would only require four credit hours (with the suggested pathways in this chapter) of remedial math. This includes 2 hours of introduction to algebra and 2 hours for the booster course

Why Should Students Who Complete Introduction to Algebra Enroll in a Booster Course?

Students need to complete or test out of an introductory algebra class before attempting a QR or introductory statistics class. However, why should students who have not completed an elementary or mid-level algebra class enroll in the booster course for QR or introduction to statistics?

> Introduction to algebra or pre-algebra gives students the basic foundational knowledge, but these students still need more help, shared Professor DeSilva.

Professor Williamson clarified:

> In the QR course, students need to be able to solve complex equations with fractions when they solve the DTI [debt-to-income ratio] problems. They need a good understanding of slope and equation of the line. These are more mid-level algebra types of topics. So, students need help with these topics as they are taking QR.

Professor Noles added:

> It's not just about learning new material [in the booster course]; it's about getting extra help and extra review. Students who have passed a basic algebra class have basic but marginal skills. They need the extra instruction, review, and practice to succeed in QR or statistics.

How Much Uniformity Is Necessary for Policies and Assessment?

When the community colleges evaluated themselves at the turn of the century, they discovered that a lack of uniformity was hindering student learning. There were shifting policies within departments regarding calculator usage as well as testing procedures. Additionally, instructors did not test the same content and differed in overall grading policies. Consequently, college administrators began requiring more uniform policies for math departments. By the end of the aughts, several faculty asserted that there was too much uniformity, which interfered with their pedagogical flexibility, subsequently hindering student learning.

What is the ideal amount of uniformity across these gatekeeper community college math classes? There needs to be a certain degree of uniformity so that student learning is not limited, but flexibility must exist for instructors to exercise pedagogical freedom. Math departments should create consistent policies regarding the use of calculators. Instructors within a department should have similar policies regarding the percentage of the final grade that is composed of proctored exams or quizzes. As noted earlier, the grading process becomes problematic when a student completes a math class where only 50% of the final grade is composed of exams and quizzes and then attempts a subsequent class where exams and quizzes are 80% of the overall grade.

Are standardized finals a good practice, especially for developmental math classes? The bottom line is that the goal of developmental math classes, as well as of college algebra and introductory statistics classes, is to prepare students for the subsequent course. A comprehensive final exam is one way to assess the knowledge that a student gained in one class. Additionally, studying for a final can help prepare the student for the following class. However, common finals can be restrictive for instructors, in that different people may phrase questions differently, and this discrepancy may confuse students. Faculty teaching the same course must collaborate to ensure they are at least covering the same content on their final exams, and this content is commonly and evenly distributed across the final exams. Furthermore, while final exams do not need to be completely standardized, instructors may want to include at least a few agreed upon questions. The key is to ensure enough standardization so that students are prepared for ensuing math classes while still allowing faculty flexibility.

Average Class Size

In the 1970s, math class sizes ranged from 30 to 100 students. In 2021, according to ThinkImpact, the average class size at American community colleges ranged between 25 and 35 students. This enrollment coincided with the math class sizes the faculty at Telford, Griffin, Sisco, Habyan, and Bordi Community Colleges reported in 2021 as well. These numbers, however, refer to traditional classes. Classes in the emporium model are much larger, with one or two tutors assisting the lead instructor.

The faculty participants mentioned that their class size for math is basically determined by their typical classroom capacity. More specifically, classrooms allotted for math classes generally do not exceed a capacity of 30–35 students. However, as time progresses, community colleges may remodel their infrastructure, and classrooms may become larger. Should average class sizes grow larger for traditional instruction in math, especially for gate-keeper courses?

> Truthfully, I would prefer to have class sizes around 20, *shared Professor Hickey*. We get such a diverse student population when it comes to ability levels and learning styles, so a smaller class size would allow me to help them more.

Professor Smith concurred:

> I feel sometimes like thirtysomething students is too many, especially when the skill level is lower for the class. I feel like I am not giving every-one as much attention as they need.

However, Professor Smith understands the rationale for average class sizes:

> Our school receives more funding for higher class sizes, so I think 30 is a good compromise. But it shouldn't be more than that.

Professor Williamson added:

> I don't want a class that's too small. I need enough students where I can engage them in discussion when teaching.

When Professors Morgan and Milacki began teaching in the 1970s, they had class sizes between 50 and 100 students. I asked them if average class sizes in traditional math classes should ever reach that level again.

> Absolutely not, *replied Professor Milacki*. I did the best I could, but I know because of the large class size, I didn't give my students the best instruction.

Professor Morgan added:

> You have to understand. Teaching that kind [gatekeeper] of math class is not like lecturing at a university. I admit, back then [the 1970s], I thought it was. You have to get to know your students. You have to understand their skill level and try to understand how to help them learn. You can't do that with too big of a class.

Additional Initiatives to Sustain Success

When planning for the future of community college math, initiatives that have proven successful in the past will likely resurface. Online and virtual learning will be part of community college math for years to come. The empo-rium model will always exist in some form. However, there are two other initiatives that will likely be part of the future.

Supplemental Instruction

Supplemental instruction (SI) has proven to positively impact students in standalone developmental math. Unlike the booster courses, SI classes allow time for students to simply review prior material without the pressure of learning new material. One or two 50-minute sessions per week work well. Three elements are essential to the success of SI: consistency, engagement, and mandatory attendance.

SI sessions are generally led by an SI leader, a tutor who has undergone training to help the students. The SI leader should attend class sessions to understand the instructor's pedagogy, as well as get a sense of the class's learning style. Consequently, the SI leader can bring consistency to the SI sessions.

SI leaders should review problems with students; however, the SI sessions are a good way to engage them. The SI leader's review should more closely resemble a group discussion than a lecture. Students can work together in groups to solve problems, which allows them to make connections they may not in class.

Lastly, SI tutorial sessions should be mandatory. The reality is that many students will not invest in opportunities that are optional. SI sessions should be built into registration so that students must register for the SI session in addition to the standalone class.

Like booster courses, SI sessions do not have the bandwidth to help all students, especially those who are very far behind in the course or lack basic prerequisite skills.

Learning Communities

Learning communities have a checkered past in community college math. In the aughts learning communities created a great deal of frustration and confusion for students. More specifically, students were frustrated by the abstract nature of learning communities, which required students to connect themes between classes and made heavy use of inquiry-based instruction. Additionally, students' schedules made arranging learning communities problematic. However, there are data that suggest that there is a strong correlation between students making connections with their peers and staying in school. Establishing meaningful connections also helped students master their math. Consequently, learning communities are likely to resurface in community college math.

Community college students can be positively impacted by learning communities in math, although convincing students to co-register for common classes will continue to be an issue, considering their busy lives. However, the key is to keep it simple. Having students co-register for two classes to form a cohort can help them form a sense of community. Additionally, instructors can help by engaging students in class using various types of collaborative learning. However, there is no need to complicate the simple idea of learning communities with higher order concepts such as identifying common themes between two different classes and mandating inquiry-based instruction.

Utilizing External Resources to Assist Students

A sustainable model for future success in math is dependent not just on the selected curriculum and teaching modalities but the organization. Community colleges must have resources that help students in their endeavor. Two vital parts of the organization are effective tutorial centers and efficient academic advising.

Tutorial Services

For decades, community colleges have utilized tutorial and academic support centers to assist students. In addition to one-to-one in-person tutoring, tutorial centers mobilize various resources to serve students. From paper and pencil worksheets to VHS tapes and CD ROMS, to Internet videos and virtual tutoring, tutorial centers have evolved with the times. One characteristic has remained constant, however: students receive outside assistance in a difficult subject.

Tutorial services remain a vital part of the students' experience in community college math. Tutors must be properly trained and effective in working with students. Above all, they must be caring individuals who travel the extra mile to help struggling students. Constant communication between math departments and tutorial services is imperative so that they are working cohesively to help students. Additionally, tutorial services must remain up to date with technology to supplement students' learning processes.

Academic Advising

The importance of solid academic advising for students is immeasurable. Academic advisors work with students when they first enter community college and continue to do so as they navigate their math requirements. The path for students' success in math is not always linear, and advisors can help students reach the right path when needed.

Community college students make up a diverse population. Consequently, there is not a one size fits all model for success. There are various pathways, and as students enter community college, advisors can help direct students toward them. If students need additional help in arithmetic, what are their options, and how do they get started? Is the emporium model a good fit for them? How about the corequisite model? Should this specific student enroll in an in-person, virtual (synchronous learning), or online class (asynchronous learning)? Advisors are generally on the front line of assisting students with these questions based on their individual background and learning style. Therefore, constant communication between math faculty and academic advisors is essential.

Student Responsibility

The term student responsibility is widely used, but it has varying definitions. I am going to employ the definition for student responsibility that I have used in previous research from the Jamestown Community College syllabus in Jamestown, New York: "Student responsibility occurs when students take an active role in their learning by recognizing they are accountable for their academic success. Student responsibility is demonstrated when students make choices and take actions which lead them toward their educational goals" (Jamestown Community College, 2020). Throughout the twenty-first century, math faculty asserted that student responsibility has waned. However, students have exhibited a lack of motivation and poor habits since the 1970s. That is, a lack of student responsibility has hindered student success since the inception of community college math education.

The reality is that community colleges can design quality math models to assist a heterogeneous mix of students. Additionally, students need support. They require quality and thorough instruction along with outside resources. However, students must do their part and accept responsibility for their own learning.

Closing Equity Gaps

The twenty-first century has witnessed large amounts of minority students placing into developmental math classes. ESL students have continued to struggle in developmental math as well. How can community colleges assist these students going forward?

As mentioned earlier, many minority students enter community colleges with more deficient math backgrounds than White students. These students tend to come from neighborhoods at or below the poverty level and received an inadequate education in elementary and high school. Since math is progressive in nature, minority students tend to develop gaps at a higher rate and fall behind in math.

ESL students face challenges when attempting community college math. Language barriers can impede students' efforts to understand their professors and communicate with classmates. Furthermore, ESL students as well as Black and Hispanic students often feel apprehensive about seeking assistance for their struggles.

Community colleges should consider addressing the root causes of equity gaps moving forward. Community colleges, especially math departments, should focus on community outreach programs for inner-city schools. More importantly, such initiatives should concentrate on Black and Hispanic students who have an interest in attending community college following graduation. After assessing their current math skills, the community college should appoint student tutors and even faculty to work with students on improving their basic math skills and college readiness. This can be accomplished individually and in small groups.

Upon admission, Black and Hispanic students, as well as ESL students, need guidance. Black and Hispanic students are more likely to lack confidence in their math skills and feel that failure is imminent. ESL students often lack confidence and feel overwhelmed in a new environment. Consequently, community colleges should do their best to pair these students with advisors as well as peer and faculty mentors who can address their concerns. This will assist these students in acclimating to their math classes and the community college environment.

Lastly, math faculty need to be educated regarding the backgrounds of Black, Hispanic, and ESL students. The reality is that all students must master the same math concepts, and in many cases, they must do this to be successful in their subsequent math classes. However, understanding previous common learning experiences, backgrounds, and needs can help faculty teach these students better.

Moving Forward with Math

Mathematics will continue to be a requirement as students complete their degrees. Therefore, the subject will remain an integral part of community college education. Community college math, and even math as it existed before community colleges, has a rich past.

In the 1970s, community college math instructors started with nothing more than a textbook and a blackboard, and through the years they have developed numerous resources, initiatives, and modalities that have helped students. This change has taken place with the use of technology, collaboration with peers, professional development organizations such as AMATYC and NADE (later renamed NOSS as the National Organization for Student Success), and their own desire to help students succeed. Some initiatives have succeeded, while others have failed; some have fallen somewhere in between. As with any discipline, it is imperative to learn lessons from the past and the present to help students in the future. Generations of students will pass through math classrooms, whether they are in person, virtual, or online. Technology will evolve. However, there will be one constant: our students will always need us.

References

Jamestown Community College (2020). Student responsibility statement. https://www.sunyjcc.edu/student-experience/student-responsibilities/student-responsibility-statement#:~:text=Responsible%20students%20take%20ownership%20ofseminars%2C%20prepared%20and%20on%20time.

ThinkImpact (2021). Community college statistics. https://www.thinkimpact.com/community-college-statistics/#1-community-college-vs-university-statistics.

Appendix A: Design of the Study

Type of Data Gathered

This study, focusing on the past, present, and future of community college math, utilized two types of empirical data: descriptive and qualitative. The descriptive data consisted of a variety of success and completion rates from the community colleges that I studied. This was largely a qualitative study, as it consisted of inductive data gathered through interviews. According to Merriam (2002), this study resembles three genres of qualitative research: a basic interpretative qualitative study, various case studies, and grounded theory. This was a basic interpretative study in that I conducted this study with the purpose of gaining various perspectives of the past and present events for community college math. There were assorted case studies, as I studied the curricula, initiatives, and data for seven community colleges. Lastly, this study touched on grounded theory, as I presented several postulations and projections based on the data I gathered.

Conducting the Study

My goal was to interview roughly 25 faculty members. I wanted to speak with a variety of faculty who taught math full-time at a community college at various times from the 1970s through 2020. There were some faculty members who I specifically sought out, as I knew the time periods in which they taught. I also contacted various people in the field of developmental education to inquire if they knew people who would be available for an interview.

Several faculty members were retired at the time of the interviews; therefore, obtaining Institutional Review Board (IRB) approval was not required. I simply needed to obtain their consent. For the faculty members who are currently employed, I acquired IRB approval at their respective institutions. I also obtained IRB approval to gain raw data regarding success and completion rates as well. Additionally, most of the institutions allowed me to search through previous course catalogs to examine their archives. In some cases, a staff member was kind enough to scan and email copies of previous course catalogs and course sequencing.

Confidentiality

This study revealed sensitive information; therefore, I took painstaking steps to ensure confidentiality. I assigned each of the faculty participants and the community colleges pseudonyms. Moreover, any person mentioned in this study was given a pseudonym. I did not even reference geographic location regarding the community colleges. To further protect the institutions' confidentiality, I configured my own figures and tables based on the raw data I was given.

Interview Questions

My sole source of qualitative data collection was interviews. To gain some basic background information on the participants, I asked them to complete a short demographics questionnaire. Then, according to Kvale and Brinkmann (2009), I utilized three types of interview questions. I began with standardized lead questions. From there, I employed follow-up questions, probes, and specifying questions to gain a deeper perspective. For all the interviews, I gathered the richest data through the aforementioned four types of questions. After interviewing 25 participants, I decided to add to the research by interviewing five more participants. As I got closer to 30 participants, I realized that I was learning less information and moving toward saturation, which is where the researcher is no longer learning new information (Krathwohl, 2009). I conducted all initial interviews via Zoom. Additionally, I conducted the, follow-up questions using either Zoom or email.

Ensuring Trustworthiness

A solid qualitative study establishes trustworthiness with the reader. I wanted to ensure that my research instrument, which was my interview questions, validated my research purpose. Therefore, I consulted with various peers who reviewed my questions and offered suggestions. Additionally, I asked my peers to review my questions for the institutions' analytics departments, so that I was obtaining valid descriptive data. Krathwohl (2009) refers to this process as peer checking.

Much of the findings from this study were based on the elaborate interviews with the participants. In an effort, to help the reader feel part of the various conversations, I utilized the raw data in the form of direct interview responses. In qualitative research, academics refer to this as a thick, rich description (Patton, 2002; Ridenour & Newman, 2008).

Analyzing the Data

After having all the interviews transcribed, I engaged in a process entitled constant comparison (Merriam, 2002). This is when the researcher reads the data numerous times with the goal of identifying common themes and eventually creating various categories.

References

Krathwohl, D. R. (2009). *Methods of Education and Social Science Research* (3rd ed.). Long Grove, IL: Waveland Press, Inc.

Kvale, S., & Brinkmann, S. (2009). *Interviews: Learning the Craft of Qualitative Research Interviewing* (2nd ed.). Thousand Oaks, CA: Sage Publications, Inc.

Merriam, S. B. (2002). Assessing and evaluating qualitative research. In. S. B. Merriam and Associates (Eds.), *Qualitative Research in Practice: Examples for Discussion and Analysis* (18–33). San Francisco, CA: Jossey-Bass.

Patton, M. Q. (2002). *Qualitative Research & Evaluation Methods* (3rd ed.). Thousand Oaks, CA: Sage Publications, Inc.

Ridenour, C., & Newman, I. (2008). *Mixed Methods Research: Exploring the Interactive Continuum*. Carbondale, IL: Southern Illinois University Press.

Appendix B: Lead Questions for Participants (1970s–2009)

1. When you first started teaching community college math, what was the modality of instruction (face-to-face, etc.)?

2. In the 1970s, did your school use a locally developed placement exam? If so, was this developed by faculty? Was the placement test graded by hand?

3. When your school went to nationally employed placement exams in the 1980s, did you notice any difference regarding more accurate placement in math courses?

4. Do you feel that the demographics (age, race, etc.) of the student population was different from when you started teaching math as opposed to the 2000s or even the present day? How so?

5. Do you feel that students were more or less motivated to learn and succeed when you first started teaching community college math as opposed to the 2000s or even present day?

6. What was your primary form of instructional delivery in the 1970s (e.g., lecture)? Did this change over time? How so?

7. Are there teaching methods that you employed prior to 2000 that you still employ? Are there methods that you employed that were successful but abandoned at some point?

8. When you first started teaching, was college algebra the course that most or all students needed to complete for their degree and to transfer to a four-year school? If not, what was the course(s)?

9. Prior to 2000, did your institution keep track of student success rates for community college math courses?

10. During the 2000s, community colleges began to feel increasing pressure to raise student success rates in mathematics. Was this the case prior to 2000? Please elaborate.

11. During the 2000s, administrators became increasingly involved in the day-to-day instruction of community college math. Was this the case prior to 2000?

12. In general, what do you feel are the biggest changes regarding community college math from when you started teaching to the present day? Such changes may include but are not limited to a math class in the present day compared to a math class during your inaugural years.

Appendix C: Lead Questions for the Participants (2000–2019)

1. Can you tell me about the sequencing of classes, from developmental math to college algebra, when you started teaching? Did this change much? If yes, how so?

2. What were the primary modalities of instruction in your institution during the 2000s and the 2010s. Can you tell me how these modalities changed over time?

3. During the 2000s, there were several initiatives to improve community college math. Can you tell me how this impacted your institution?

4. Developmental math, in particular, got bombarded with statistics, mostly negative statistics, in the early 2000s. Was this the case at your institution? How did this drive your course sequencing and instruction?

5. How was the relationship between your department and the administration during the 2000s? Did you feel supported in your teaching?

6. How did your institution handle arithmetic courses in the 2000s?

7. Please tell me about the student population in the 2000s? What were the challenges?

8. Did distance learning continue to grow at your school?

9. During the 2000s, one issue that arose in community college math was the high number of Blacks and Hispanics who placed into developmental math. Did you notice this at your institution?

10. Please tell me about the uniformity in your department. Was there a struggle to get faculty to utilize similar policies regarding grading and calculator usage?

11. Introduction to statistics and college algebra did not receive too much attention during the 2000s (first decade of the 21st century). Can you tell me about these courses during that time period?

12. During the 2010s, many institutions began implementing alternative math pathways with corequisites. Did your school do this, and how did this go?

13. Some schools have discussed or even eliminated standalone developmental math. How do you feel about this?

14. During the 2010s, higher education accrediting bodies cracked down on the qualifications for teaching college-level math. Did this impact you in any way?

Appendix D: Lead Questions for the Participants (during the Pandemic and Beyond)

1. When your institution got shut down, due to COVID-19, how did your department handle instruction?
2. What were some of the challenges you faced with virtual instruction?
3. How did your institution handle remote testing?
4. Institutions have different ways of handling students who struggle in arithmetic. How did this change during the pandemic?
5. What were some of the positive outcomes of virtual instruction?
6. What place does virtual instruction have in the future of community college math?
7. Booster courses seem to be in the future of community college math. What is the best way to utilize them?
8. In your opinion, what is an ideal community college math sequence, especially for gatekeeper courses? What is the best way to ensure that students are sufficiently prepared but that they also do not get lost in math pathways?

Cumulative Interview Notes (Appendices B–D)

- In many cases, the participants' responses from the lead questions led to several follow-up questions, probes, and specified questions.
- Oftentimes, responses from various participants led to follow-up questions for participants who I previously interviewed.
- Depending on the participant's background, certain questions were omitted. For example, I did not ask faculty who already taught college-level math and possessed the proper credentialing about the mandate from the higher education accrediting bodies.

Appendix E: Demographic Questions

1. What year did you start teaching community college math?
2. During your first few years of teaching, which courses did you teach?
3. Have you taught developmental math courses? When?
4. What are your degrees and majors for each degree?

Index

Note: **Bold** page numbers refer to tables; *italic* page numbers refer to figures.

abysmal skills 104
academic advising 232–233
academic counselors 68
Academic Quality Improvement
      Program (AQIP) 98, 99
Accelerated Learning Model 150
acceleration 89, 125
accreditation credentialing issues
    136–138
ACCUPLACER 42, 55, 134, 140, 162
ACT scores 141
Achieving the Dream 88, 89, 90, 93, 94,
    96, 100
admissions standards, national level
    108–109
adult basic education (ABE) 132–135, 151,
    174, 190
age demographics 73
ALEKS software 134, 171, 176
algebra 2, 7, 18, 153
    basic concepts 86, 87, 114
    booster course 228
alternative Math pathways 148, 163,
    204–205, 213
    corequisites 149, 150
    Quantway 149
    Statway 148, *149*
    time and cost for students
        227–228
ambiguity 208
American community college 9, 165
    system 1, 8
American Educational Research
    Association 139
American higher education 32
    inception of 1–2
    system 2
American Mathematical Association of
    Two-Year Colleges (AMATYC)
    15, 16, 18, 26, 27, 36, 43, 45, 52,
    53, 63, 71, 72
American Missionary Society 4

Apple II computers 54
arithmetic classes 128, 189–190, 201
    decline of 131–135, 137
    success rates in 201
arithmetic concepts 133, 160–161
arithmetic courses, higher demand
    104–105
arithmetic refresher course 133–135, 151,
    174, 217
arithmetic skills 201, 217
aughts 81–117, 202, 203
Avenoso, Frank 15
average class size 229–230

Ballard (Professor) 13, 18, 23–24, 26–27,
    29–31, 41–43, 46, 47, 49, 50, 52,
    53, 63, 64, 66, 67, 72–74, 76, 78,
    96, 98–100, 104, 106, 113, 201
basic calculator 53; *see also* calculators
BCC *see* Bordi Community College
    (BCC)
behavioral issues 106–107
Bell (Professor) 35, 44, 45, 48, 52–55, 73,
    76, 95, 107, 109–112, 115, 116, 131,
    132, 135, 136, 139, 153, 160, 165,
    167, 172
Bill and Melinda Gates Foundation 123
Black and Hispanic students 169,
    170, 172
Bloom's taxonomy 208, *209*
booster courses 152, 156, 162, 206
booster courses, logistics 225
    instruction 227
    length 225–226
bootcamps 126, 127
Bordi Community College (BCC) 14,
    20–21, 32, 38–39, 48, 50, 53, 62,
    65, 82, 90, 121, 129, 134, 147, 150,
    157, 182, 183, 187
    basic concepts of arithmetic 38
    basic concepts of Mathematics 20
    college algebra 21

Bordi Community College (BCC) (*cont.*)
 fundamentals of algebra 1 20–21, 39
 fundamentals of algebra 2 20–21, 39
 fundamentals of algebra 3 39
 pressure to compress, quarter to
  semester system *127*, 127–128
 standalone developmental Math
  162–163
 statistics 21
Brown (Professor) 182, 183, 185, 188, 191,
  194, 196
Brown, Joseph Stanley 5

calculators 52–53, 80, 112
 policies 30, 70–73
calculus 7, 136, 138
California Community Colleges 83
Carnegie Foundation for Teaching
  Advancement 148, 149
Casio t (TI) 72
CCA *see* Complete College America
  (CCA)
Charles A. Dana Center 149
Chesapeake Community College 16
civil rights movements 7
collaboration 111–112
collaborative learning 95
college advisors 174
college algebra 116, 123, 125, 139, 140, 148,
  150, 152, 153, 156, 157, 159–160,
  200, 211, 224
 booster course 224–225
 and statistics 114
College Entrance Examination Board
  (CEEB) 4
college-level course 152, 158, 159
college-level Math 137, 141, 150, 157
college Math curriculum 2, 200
Colorado Community College system 83
Common Core 142–143, *143*
Community College, rise of 7–8
Community College Math
 student population 27–28
 virtual and online 181–182
Community College model, uniqueness
  8–9
Community College of Baltimore 150
Compass 55

Complete College America (CCA)
  123–124, 205, 206
compression
 quarter system to semester
  system 124
 Bordi Community College *127*,
  127–128
 Sisco Community College
  125–127, *126*
 Telford Community College
  128, *129*
computational skills 125, 137
computer adaptive placement
  exams 140
computer software, emergence 54
cooperative learning 4
corequisites 149–152, 156
 quantitative reasoning with 153–155
 Statway with 152–153
Coronavirus pandemic 181, 197
cost factors 203
cost reduction, open educational
  resources 176–178
course restructure
 persistence rates 84–85
 reversed 87–88
course structure 37, 59
 basic concepts of arithmetic I 59
 basic concepts of arithmetic II 59
 Bordi Community College 20–21,
  38–39
 changes in 41–42
 Griffin Community College 22, 40
 Habyan Community College 21, 39
 Lester Community College 22–23, 40
 Sisco Community College 19, 37–38
 Telford Community College 20, 38
critical thinking 149
Cuisenaire rods 76
cultural audit 203
culture change 201–203
culture shock 201–203

Dana Center and Carnegie 153
DeLeon (Professor) 14–15, 17, 18, 24, 28,
  30, 31, 43, 44, 46, 48, 50, 62, 67,
  68, 71, 73, 75, 86, 87
department counselors 68

DeSilva (Professor) 121, 125, 126, 132,
134, 135, 140, 141, 150, 153–155,
169, 171–176, 183, 189–191, 228
developmental course sequence 125, 127,
128, 206
developmental education 1, 6, 123, 124,
137, 199
developmental Math 6–7, 24, 31, 89, 108,
124–131, 137, 142, 143, 149–151,
154, 165, 167–169, 171, 202; *see
also* standalone developmental
Math
classes 167, 168
and Community College Math
199–200
reform effort 117
statistics and 83
Developmental Mathematics
Curriculum Committee 16
DEV Math classes 129, 130, 137, 170, 171
Dewey, John 98
distance learning 181, 207
and online learning 102–103
Douglass (Professor) 123, 126, 127, 135,
141, 151, 172, 174, 196, 197
dual enrollment 138–140

Edline system 66
educating faculty, ESL students 172–173
effective instructors 208
effective teachers 208
effective teaching 208–209
elementary algebra 123, 128, 153, 219–220
email communication 67
employment 3
emporium model 51, 79, 89–92, 97, 128,
129–131, 186, 187, 206–207
updated 90
engagement 208
lack of 184–186
English as a Second Language (ESL)
students 164–165, 168, 234
additional resources for 166–167
challenges 169
challenges for 75–76
educating faculty on 172–173
rise of 75
understanding of 165–166

enrollment flexibility 227
equity concerns 105–106
equity gaps 233–234
for minority students, Math 169
Euclidean geometry 2
exams 110–111
explanations 193–194
external resources 232

face-to-face instruction 61, 134, 181, 193,
207, 212, 219
faculty and administrators, relations
77–78
faculty credentialing issues 136–138
Fenimore (Professor) 15, 24, 28, 31, 46, 48,
51, 64, 65, 85, 88, 110
financial aid system 135
Florida, developmental education
reform 157
formal department meetings 30–31
free software 176–178
full-length arithmetic courses 200–201
full-time equivalent (FTE) 8, 84

gatekeeper Math courses 143
Gates Foundation 89, 130, 131
General Education classes 161
general Math class 155–156
general Mathematics 60
Georgia, standalone developmental
Math elimination 157
G.I. Bill of Rights 6
good practices 205–206
Google Hangouts 183
grading 140
policies 112–113
uniformity 113
graphing calculator 72, 73
Great Depression 6
Greek verbs 1
Griffin Community College (GCC)
14–15, 17, 18, 22, 35, 40, 47, 53, 57,
59, 64, 68, 71, 82, 86–87, 96, 111,
114, 115, 122, 130, 133, 148, 157,
158, 187, 211
algebra 1 and 2 22, 40
basic arithmetic 22, 40
college algebra with trigonometry 22

Griffin Community College (GCC) (*cont.*)
  early intervention program at 175
  statistics 22
  STEM and non-STEM students, Math
    pathways 163, *163–164*
Gross, Herb 16
group-based instruction 154, 155–156
group-work 4
Guzman (Professor) 121–122, 128–129,
    131, 132, 134, 139, 142, 143, 151,
    154–158, 163, 175, 177, 183, 186,
    192, 194, 226

Habyan Community College (HCC) 14,
    16, 21, 25, 35, 39, 44, 53, 59, 70,
    93, 95, 101, 122, 131, 132, 160
  basic Mathematics 21, 39
  college algebra 21
  elementary algebra 21, 39
  intermediate algebra 21, 39
  introduction to algebra 39
  standalone developmental Math.
    162–163
  statistics 21
Harnich (Professor) 147, 151, 158, 161,
    163, 176, 177, 185, 192, 195, 196
Harper, William Rainey 5, 140
Harvard, Reverend John 1
Harvard College 1
Hewlett Packard 30
Hickey (Professor) 122, 130, 132, 152, 153,
    156, 161, 163, 166–168, 172, 176,
    177, 188, 190, 191, 212, 225
higher comfort level 190–191
higher education 1–10
  changing landscape of 2–4
  start of Mathematics in 2
Higher Education Act (HEA) 7
higher-level Math classes 136
higher-lever ESL Students 165
Holton (Professor) 122, 130, 132–134, 138,
    161, 162, 184, 191, 192
Housatonic Community College 83
hybrid emporium model 130

Indiana Community College system 83
inner-city communities, outreach
    170–171

inner-city high school 170
in-person learning 207
in-person testing 187–189
inquiry-based instruction 96, 97
inquiry method 153, 154
instruction, method 50–51
instructional modalities 61–67
  contextualization 62–63
  distance learning development
    63–67
  emporium model, phasing
    out 62
  lecture plus 61
  for Math courses 23–26
instructor absences 195–196
intermediate algebra 123, 127, 150,
    156, 220–221
Internet videos 175–176

Johnson (Professor) 36, 51, 54, 62–64,
    73, 91, 103, 131
Johnson, Lyndon 7
Joliet Township High School 5
Jones, Stan 123
junior colleges 9

K–12 education 136, 142
Khan, Sal 175
Khan Academy 175
Kilgus Community College 36, 51,
    59, 62, 64, 73, 91, 131

laissez-faire approach 17
laissez-faire policy 18
Land Grant institutions 9
Latin poetry 1
LBCC *see* Long Beach City College
    (LBCC)
LCC *see* Lester Community College
    (LCC)
learning communities 25, 26, 95, 100,
    231–232
  return of 93–95
learning disabilities 29–30
lecture 95–98
lecture plus 61
legitimate faculty buy-in 209–210
less anxiety 190–191

Lester Community College (LCC) 15,
22–24, 40, 43, 48, 51, 59, 62, 64,
68, 85, 124, 201
college algebra 23
elementary algebra 40
fundamentals of arithmetic 40
fundamentals of Mathematics 22
fundamentals of Mathematics I and
II 22–23
intermediate algebra 40
introduction to Mathematics 40
statistics 23
Limited English Proficiency (LEP) 165, 166
Long Beach City College (LBCC) 141
Lopez (Professor) 82, 90–92, 108, 109, 129,
136, 158
lower *vs.* higher-level Math students
28–29

Maricopa Community College 83
Martin, Deanna 100
Massey, John 16
Mathematical American Association
(MAA) 4, 6
Mathematical maturity 153
Math faculty, professional development
167–168
Math software programs 207
Math tutorials, development 67–68
Math wars 78–79
McDonald (Professor) 57, 61, 66, 74, 92,
106, 108, 133, 143, 162, 175, 183
Meiklejohn, Alexander 26
Mesa (Professor) 57, 61, 66–68, 75, 77, 81,
94–95, 101, 102, 106, 107, 114, 125,
128, 132, 134, 135, 137, 165, 166,
169, 171, 177, 183, 188, 189
messages, conflicting 109–110
Milacki (Professor) 14, 16, 17, 23, 25, 27,
28, 30, 31, 45, 46, 48, 50, 51, 53,
60, 65, 67, 69, 74, 77, 90–92, 106,
111–113, 230
Miller, George 15
minimesters 126, 127
minority students 169–170
assistance and guidance for 172
in Math 169
outreach program for 170–171

Mitchell (Professor) 14, 17, 23, 31, 32,
43, 48–50, 55, 59, 71, 74, 78, 81,
86–88, 100–102, 104, 111, 114,
116, 136
MOM (My Open Math) 177
money 203–204
Morgan (Professor) 14, 16, 17, 25–30,
32, 41, 42, 44, 53, 54, 63, 67, 70,
73–75, 77, 78, 93–94, 101, 104,
106, 107, 112, 125, 230
Morrill Land Grant Act 3, 7
Moyer (Professor) 82, 110, 112, 116, 125,
128, 132, 134, 136, 140, 152,
155, 156, 169, 171–174, 177, 190,
194, 226
multiple measures 140–142
Mussina (Professor) 82, 90, 91, 104, 109,
115, 116, 129, 156–158, 161, 163,
176, 186, 212
My Lab Math/MyMath Lab 91, 92, 102,
103, 176, 177, 183

NADE Self-evaluation Guides 58
National Association for Developmental
Education (NADE) 36–37
National Center for Academic Transition
128, 133
National Council of Teachers of
Mathematics (NCTM) 79
National Defense Education Act 6
National Governors Association 142
Networked Improvement Community
(NIC) 148
NIC *see* Networked Improvement
Community (NIC)
Noles (Professor) 122, 128, 133, 134, 141,
152–156, 168–171, 175, 183, 188,
193, 195, 212, 227, 228
non-credentialed faculty, standalone
developmental Math
elimination 162

online graduate-level Math classes 138
online instruction 207
online learning 102–103, 182, 207
open educational resources 176–178
organizational culture 202
types of 202

participants 182
  additional 82
  faculty members 13
  returning 81–82, 215
peer mentoring program 166
Pell grants 7
persistence rates 86
Pima Community College 83
placement exams 55
placement testing 16–18
positive relationships, administration
    54–55
power-coercive approach 210
practice tests 27
pre-algebra 218–219
preparation, quality instruction 45–46
pressure to compress 124–125
  Bordi Community College *127*,
    127–128
  Sisco Community College 125–127, *126*
  Telford Community College 128, *129*
Price University 138
primary instructional method 78
professional development 15–18
  AMATYC (American Mathematical
    Association of Two-Year
    Colleges) 36, 58
  NADE (National Association for
    Developmental Education)
    36–37, 58–59
progressive movement 4–5
public community colleges 7

QR booster 150
QR course 211
quantitative reasoning (QR) 149, *149*,
    150, 151, 158, 204, 205, 216, 219,
    221, 228
  benefits of 155–156
  booster course 221–222
  with corequisites 153–155
Quantway 149, 150
  applications 150–152, *152*
    arithmetic score 160–161
    QR benefits 155–156
    QR with corequisites 153–155
    standalone developmental Math
      elimination 157–162, **159, 160**

recorded sessions 192–193
relationships, administration 31–32
remedial education 3
remote learning 195
remote testing 189
remote tutoring 197
restructure, rationale 60
Rhodes (Professor) 182–185, 189, 191, 192,
    194, 225

SAT scores 141
School Mathematics Study Group 6
scientific calculators 52, 194
screencast 184
Screencast-O-Matic 184
second Morrill Land Grant Act 3, 4
*Section 504 of the Rehabilitation Act of
    1973*, United States 29
service-learning 98–100
Serviceman's Readjustment Act 6
shorter attention spans 107
Sisco Community College (SCC) 13, 18,
    24, 37–38, 59, 72, 82, 98, 102, 121,
    123, 150, 173, 201
  algebra 1 and 2 19, 37–38
  algebra 3 37–38
  basic Mathematics 19, 37
  college algebra 19
  multiple measures 141
  pressure to compress, quarter
    system to semester system
    125–127, *126*
  statistics 19
skill level 16
skills course 49
Smith (Professor) 148, 158–160, 184, 191,
    193, 195, 211, 226, 230
social distancing 197
social issues 106–107
software programs 176
Sputnik 6, 9
standalone developmental Math
  elimination 157–158, 161–162
    impact on non-credentialed
      faculty 162
    increase completion rates 158–160,
      159, 160
standardized arithmetic exam 2

standardized placement testing 42–43
state funding, success rates 84
state legislatures 124
statistics 211, 222–223
   Twenty-First century 115–116
   booster course 152, 153, 223–224
   and college algebra 114
   developmental Math and 83
Statway 148, *149*, 150
   applications 150–152, *152*
      arithmetic score 160–161
      QR benefits 155–156
      QR with corequisites 153–155
      standalone developmental Math
         elimination 157–162, **159, 160**
   with corequisites 152–153, 156
Stephens (Professor) 182, 186, 193,
     195, 197
straight emporium model 131
student consumerism 210–211
student engagement, improving 43–45
student entitlement 107–108, 210–211
student outreach 68–69, 170–171,
     173, 189
student placement 141
student population
   ages and changes 73–74
   challenging 103
   changing 46–47
student profiling 173–174
   early intervention 175
student responsibility 233
students, counseling 49–50
student success in Math improvement,
     national initiatives 88–89
study guides 27
study skills, emphasis 48–49
supplemental instruction (SI)
     100–102, 231
sustain success
   additional initiatives 230
   pathways 215–216, *217*
Sutcliffe (Professor) 82, 87, 91, 97, 102,
     104, 108, 112, 125, 130, 131, 137,
     138, 161, 162, 175, 188, 192, 195

take-home exams 187
teacher preparatory classes 76–77

technology, Community College Math
     Courses 52
Telford Community College (TCC)
     14, 38, 49, 57–60, 71, 82, 86, 95,
     100–102, 114, 122, 133, 155, 173,
     182, 188, 197, 212
   basic concepts of algebra 38
   college advisors at 174
   college algebra 20
   computational skills 38
   elementary algebra 38
   ESL students in 165–166
   fundamentals of Mathematics 20
   intermediate algebra 20, 38
   minority students in 169
   multiple measures 141
   pressure to compress, quarter system
     to semester system 128, *129*
   statistics 20
testing, in-person 187–189
Thurmond (Professor) 35, 47, 53, 54, 64,
     65, 70, 74, 77, 81, 87, 96, 97, 104–
     107, 111, 112, 130, 133, 137, 162
TI-83 graphing calculator 72
Timlin (Professor) 57, 72, 73, 75, 76, 87,
     112, 115, 136, 140, 143, 175
Transfer Modules 134, 137, 154–155
trigonometry 2
Trombley (Professor) 82, 102, 108, 110,
     135, 136, 142, 154, 170–172, 174
tutorial services 232
   birth of 47–48
tutoring 1
tutors 1, 3
Two-Year Junior College 5

uniformity 113–114
   lack of 110
   policies and assessment 228–229
"universal Mathematics" 6
University of Chicago 5
University of Connecticut 139
University of Wisconsin 3, 26
U.S. Department of Higher Education
     131–132, 135, 143–144

VHS tapes 53, 55, 63, 64
video conferencing 183, 197

videos 53
Virginia Community College system 83
virtual booster classes 226–227
virtual classes 195, 196
    style and role of 194–195
virtual instruction 190, 191, 207
    accessibility for students 195
    conversion 183–184
    success rates for 193
virtual learning 181, 197
virtual tutoring 196–197
volume of questions 186

Wallace (Professor) 15, 23, 30, 42, 43,
    68, 201

whole numbers 218
Williamson (Professor) 122, 131, 133,
    139, 152, 154–156, 161, 166,
    167, 177, 184, 185, 189, 190, 194,
    226–228, 230
W.K. Kellogg Foundation at
    Appalachian State
    University 36
Women in Transition (WIT) program
    69–70
World War II 9

YouTube videos 175

Zoom 183, 185, 186, 188, 191, 192, 196

Printed in the United States
by Baker & Taylor Publisher Services

Printed in the United States
by Baker & Taylor Publisher Services